Palaeochannels and Water Resource Management

Sadik Mahammad • Aznarul Islam

Palaeochannels and Water Resource Management

A Study in the Western Bengal Delta

Springer

Sadik Mahammad
Department of Geography
Aliah University
Kolkata, West Bengal, India

Aznarul Islam
Department of Geography
Aliah University
Kolkata, West Bengal, India

ISBN 978-3-031-92020-2 ISBN 978-3-031-92021-9 (eBook)
https://doi.org/10.1007/978-3-031-92021-9

This Springer imprint is published by the registered company Springer Nature Switzerland AG
The registered company address is: Gewerbestrasse 11, 6330 Cham, Switzerland

Panoramic view of a palaeochannel as an abandoned channel in the Damodar Fan Delta, West Bengal, India (Field Photograph, 2021)

This book is dedicated to the victims of severe flooding and water-stress conditions in the western part of the Damodar Fan Delta, whose vulnerability can be reduced through sustainable water resource management using palaeochannels.

Foreword

Geomorphology is a dynamic and widely applied science that has progressed over the last two decades in leaps and bounds, hand in hand with the rapid advancement of new geophysical, remote sensing, and landform dating techniques. Landforms and surface processes are at the core of physical geomorphology and are critical to the understanding of the climatic changes and developing strategies to meet the future requirements of the society (Anderson & Anderson, 2010). In this context, the book *Palaeochannels and Water Resource Management: A Study in the Western Bengal Delta* by Sadik Mahammad and Aznarul Islam is an important addition to the scientific knowledge base.

The megafans or inland deltas, as they are often referred to, are an important landform. Such landforms in the Himalayan foothills (e.g., Kosi River 'inland delta'; c.f. Geddes, 1960) have long attracted the attention of earth scientists. But similar landforms in the western part of the Bengal and elsewhere (e.g., the Damodar megafan) have remained much less investigated. Additionally, western part of the West Bengal, unlike its eastern part fed by Himalayan rivers, is fed principally by inland drainage system. Thus, water resources are scarce in these areas, and the area investigated by the authors is continually subjected to alternating periods of flood and draught. This book by Sadik Mahammad and Aznarul Islam will fill a long-felt need for investigation of the geomorphology and channel system of the Damodar fan, with the aim of solving water resource management problems in the area.

One of the unique features of this book is its emphasis on the identification and analysis of the paleochannels on the Damodar megafan and relating these paleochannels to the sustainable development of ground water. I believe this book will be immensely helpful to students and young professionals involved in groundwater monitoring and management, as well as those studying paleochannels. I congratulate the authors for their significant contribution to the investigation of the Damodar megafan and developing new approaches for groundwater resource management using paleochannels.

Tapan Chakraborty
Geological Studies Unit
Indian Statistical Institute
Kolkata, India

References

Anderson, R. S. & Anderson, S. P. (2010) *Geomorphology: The mechanics and chemistry of landscapes* (637 pp). Cambridge University Press.

Geddes, A. (1960). The alluvial morphology of the Indo-Gangetic plain: Its mapping and geographical significance. *Institute of British Geographers Transactions and Papers, 28*, 253–276.

Preface

Water is vital for the survival of all organisms, including human beings, and is essential for maintaining environmental balance. The distribution, form, and availability of water resources are not uniform across the earth. The acceptable quality and adequate quantity of water resources can minimize the cost of water treatment for drinking and industrial purposes and improve agricultural production significantly. The rapid growth of population, expansion of agricultural activities, and increasing urbanization escalate the demands of water for domestic, irrigation, as well as industrial uses, which may accelerate the depletion of water resource in terms of quality as well as quantity. In addition, the high variability of climatic parameters has a significant influence on water resources, which causes social and economic insecurity in the livelihoods of local people. Floods and drought are common hydro-meteorological phenomena that occur due to abnormal behaviour of precipitation. Besides, anthropogenic activities near the fluvial system alter the morphological parameters of the river, causing congested flow of water and increasing siltation in the riverbeds, which may be responsible for flooding. Moreover, characteristics such as geology, vegetation, and soil types and human activities such as overexploitation of water influence the propagation of drought conditions. Therefore, planning, monitoring, development, and management of water resources are required to deal with the surplus and deficit conditions of the water. This requires the integration of principles of earth sciences, geography, environmental sciences, and geospatial sciences to comprehend the underlying principles governing the process and outcomes of the surface and groundwater utilizations at the global and regional scale in the wake of climate change and anthropocene. Thus, this book is committed to uncovering these perspectives from an integrated outlook.

The potential of the palaeochannel in the field of water resource management by reducing the excess monsoon flood and augmenting rainwater for groundwater restoration is virtually unexplored in available books. Thus, there is a long demand for a book on palaeochannel and its potential for water resource management from a systematic approach that would help not only the regional context but also cater to the need of international demand. Thus, this book is expected to fulfil the gaps in the knowledge of palaeochannels and sustainable water resource management. The

Bengal Delta, the most populated region in the world, exerts huge pressure on land. However, there is no single book published to date on the water resource management of surplus and deficit regions through the palaeochannels of the Bengal Delta. This book intends to fill out the long-felt gap in this region. Therefore, this book will not only cater to the needs of the students at the undergraduate and graduate level in earth sciences, geography, geospatial sciences, and environmental sciences but also fill up the gap of the regional planners and decision-makers in comprehending the complexity of the Bengal Delta.

We are very much thankful to the respondents during the field survey and other stakeholders directly and indirectly associated with the data collection process. We kindly acknowledge the active and overwhelming cooperation of our laboratory members and students, including Dr. Biplab Sarkar, Susmita Ghosh, Dr. Mofizul Hoque, Abujar Rahaman, Ratnadeep Chanda, Fahad Hossain Molla, Souradeep De, Babusona Sekh, Sahajahan Rahaman, and Dipankar Ruidas for data collection and preparing some maps and diagrams. We are also equally thankful to our family members for their daily support and sacrifices throughout the entire journey. Moreover, as a larger part of this book is from the PhD thesis of Dr. Sadik Mahammad, all the concerned persons from Aliah University, Kolkata and other institutions and organizations are duly acknowledged. We also show our deep gratitude to all the other persons who promptly replied to our request for study materials and data support. We are grateful to the officials of Central Ground Water Board, Kolkata, Government of India, the Public Health Engineering Department (Head Office, Kolkata), Hooghly, Barddhaman, and Howrah District Division Office, State Water Investigation Department, Kolkata, Government of West Bengal, for providing the valuable datasets.

We are grateful to ppa. Christoph Baumann, Editorial Director of Asset Management at Springer Nature Switzerland AG, Aaron Schiller, Associate Editor, and Dr. Robert Doe, Editor of Earth Sciences, Geography, and Environment, for signing the agreement on this project and for providing us with the opportunity to publish our work in their prestigious pages. We are also grateful to the production team for their meticulous efforts in publishing this book as accurately as possible.

We hope that the outcome of this volume will ignite the minds of the scholars, faculties, planners, and stakeholders for sustainable water resource management using palaeochannels.

Kolkata, West Bengal, India — Sadik Mahammad
Kolkata, West Bengal, India — Aznarul Islam

Conversion Table

Area

1 Hectare = 2.4711 acre
1 Acre = 0.4047 hectare
1 Acre = 3.025 bigha
1 Hectare = 7.4751 bigha
1 Acre = 4047 sq. metre
1 Acre=100 satak
1 Acre = 43560 sq. ft
1 Bigha= 20 kotha

Weight

1 Maund = 37.3242 kg
1 Quintal = 100 kilogram
1 Quintal = 2.6792 maund
1 Maund = 40 kg

Competing Interests The authors have no competing interests to declare that are relevant to the content of this manuscript.

Contents

About the Book

The severity of water stress condition and concern of the people in Damodar Fan-Delta, India (Field photo, 2021).

Palaeochannels are remnants of the naturally evolving fluvial systems that have the potential for water resource management. However, the deterioration of the palaeochannels due to human interventions (urbanization, agricultural expansion, etc.) is widely noted worldwide. Linking palaeochannels to water resource management could be one of the viable options in the era of the Anthropocene and climate change when hydrological extremities in the form of floods and droughts have been exacerbated worldwide. Moreover, the rapid growth of population, expansion of agricultural activities, and increasing urbanization escalate the demands of water for domestic, irrigation, as well as industrial uses, which may accelerate the depletion of groundwater levels and deteriorate its quality in the largest part of the Bengal Delta. Besides, floods and drought are common hydro-meteorological phenomena that occur due to abnormal behaviour of precipitation. Besides, anthropogenic activities near the fluvial system alter the morphological parameters of the river, causing congested flow of water and increasing siltation in the riverbeds, which may be responsible for flooding. Thus, these extreme hydrological characteristics (surplus and deficit conditions of the water) of the Bengal Delta attracted researchers and planners to be inquisitive about these phenomena and their remedies for socio-economic development. This work intends to examine the nature, evolution, and characterization of the palaeochannels that may be crucial for sustainable water resource management.

Low-cost geospatial identifications of the palaeochannels coupled with geophysical exploration will be attempted in this book. Application of machine learning algorithms and field data will exhibit the nature (e.g. recharge potential) and evolution pattern of the palaeochannels, especially in the western part of the Bengal Delta. Some exciting case studies will be presented from the Damodar Fan Delta (DFD). The DFD is unique because once it was called as the 'Sorrow of Bengal' owing to colossal floods. Moreover, it portrays the water-stressed conditions in the northern part, especially in the lean months (January–May). This book also demonstrates how the volumetric and qualitative increase in both the surface and groundwater resources could be achieved through palaeochannels. Thus, the main objective of this work is to promote the restoration of the palaeochannels which will not only maintain the ecological balance of nature but also foster integration between palaeochannels and water resource management for society, economy, and ecology.

About the Authors

Sadik Mahammad completed his master's degree (M.A.) and PhD in Geography from Aliah University, Kolkata, West Bengal. He qualified for UGC NET JRF and WBCSC SET in Geography. Presently, he is working as an Assistant Teacher in a Higher Secondary School, under the Government of West Bengal. Previously, he was a Senior Research Fellow in Aliah University and served as a Guest Faculty in the Department of Geography, Jadavpur University, Kolkata, West Bengal. His PhD topic is 'Palaeochannels and Water Resource Management of Damodar Fan–Delta, West Bengal'. He already published 12 research articles in different international journals. He also published 5 book chapters as a principal author in international journals. He presented papers at various national and international seminars and conferences. His research area encompasses fluvial geomorphology, applied hydrology, palaeogeomorphology, water resource management and RS & GIS. Throughout the research work, he wishes to draw attention to the sustainable management of natural resources.

Aznarul Islam completed Master of Science in Geography from Kalyani University, West Bengal and MPhil and PhD in Geography from the University of Burdwan, West Bengal, India. He is an Assistant Professor in the Department of Geography, Aliah University (AU), Kolkata, India since 2016 and served as the Head (Officiating), Geography, AU from November 2019 to February 2022. Previously he was an Assistant Professor (West Bengal Education Service) at the Department of Geography, Barasat Government College, West Bengal (2014–2016). Dr. Islam served as the Guest Faculty/visiting faculty in a few institutes like Jadavpur University, Kolkata, and Kazi Nazrul University, Asansol. He was awarded the UGC Junior Research Fellowship in Geography in 2012. Besides, he secured All India Rank 018/188 in Earth Sciences in CSIR-UGC NET in 2011. He has already published more than 80 research papers including 50 SCI journal articles. He is the principal author of the book 'Riverbank Erosion in the Bengal Delta: An Integrated Perspective' published in 2024 by Springer. He edited another 10 books and contributed more than 26 book chapters in edited volumes and conference proceedings. He has presented papers in more than 35 national and international seminars and conferences. He has successfully guided 4 PhD theses and 2 theses under preparation. He was awarded one UGC minor research project. He completed one major research project funded by the ICSSR, Govt. of India in 2019 as the sole Project Director. He received a Department of Science and Technology (Govt. of India)-Science and Engineering Research Board (DST-SERB) State University Research Excellence (SURE) Award in 2023 as the Principal Investigator. He has been performing the role of editor and reviewer in various journals published by Springer, Elsevier, Taylor and Francis, Wiley, Nature, MDPI, Frontiers etc. He is a life member of the Indian Geographical Foundation (IGF), Kolkata, and the National Association of Geographers, India (NAGI), New Delhi. His principal area of research includes exploring river, water and society through the lens of applied spatial sciences using quantitative techniques and geoinformatics for reducing fluvial and hydrological hazards to enhance community relicense.

Abbreviations

AET	Actual evapotranspiration
AHP	Analytical hierarchical process
ANN	Artificial neural network
ANOVA	Analysis of variance
ARIMA	Autoregressive integrated moving average
As	Arsenic
ASF	Alaska Satellite Facility
BDL	Below detection limit
BIS	Bureau of Indian Standards
CA	Cluster analysis
Ca^{2+}	Calcium
CART	Classification and regression tree
CD	Community development
CF	Contamination factor
CFF	Chotanagpur Foothill Fault
CGWB	Central Groundwater Board
Cl^{-}	Chloride
CP	Compromise programming
DA	Discriminate analysis
DCS	Decorrelation stretch
DEM	Digital elevation model
DF	Damodar Fault
DFD	Damodar Fan Delta
DN	Digital number
DVC	Damodar Valley Corporation
DW	Dug well
EC	Electrical conductivity
EDL	Extreme danger level
EI	Eutrophication Index
EIWQI	Entropy-based irrigation water quality index
ESA	European Space Agency

ET	Evapotranspiration
F	Fluoride
FA	Factor analysis
FAHP	Fuzzy analytic hierarchical process
FAO	Food and Agriculture Organization
FCC	False colour composite
Fe	Iron
FFANN	Feed-forward neural network
GEC	Groundwater estimation committee
GEE	Google earth engine
GIS	Geographical information system
GloVis	Global visualization viewer
GPR	Ground-penetrating radar
GRD	Ground range detected
GT	Grand Trunk
GWDI	Groundwater Drought Index
GWE	Groundwater elevation
GWL	Groundwater level
GWQ	Groundwater quality
HCO_3^-	Bicarbonate
HPI	Heavy Metal Pollution Index
IBM	International Business Machines
IDW	Inverse distance weighted
IW	Interferometric wide
IWRM	Integrated water resource management
K^+	Potassium
KNN	K-nearest neighbour
KR	Kelley's ratio
LiDAR	Light detecting and ranging
LR	Linear regression
LULC	Land use and land cover
MAI	Moisture adequacy index
m-CD	Modified degree of contamination
MCDM	Multi-criteria decision-making
MFF	Medinipur Farakka Fault
Mg^{2+}	Magnesium
ML	Machine learning
MLR	Multivariate linear regression models
Mn	Manganese
MNDWI	Modified normalized difference water index
MODIS	Moderate resolution imaging spectroradiometer
MR	Magnesium ratio
MSE	Mean square error
MSI	Multi-spectral imaging
MSIs	Multi-spectral satellite images

MSL	Mean sea level
Na^+	Sodium
NDVI	Normalized difference vegetation index
NDWI	Normalized difference water index
NeI	Nemerow index
NIR	Near-infrared
NTH	National test house
NTU	Nephelometric turbidity unit
OA	Overall accuracy
OIP	Overall index of pollution
OLI	Operational land imager
OPI	Organic pollution index
OWA	Ordered weighted averaging
PA	Producer accuracy
PCA	Principal component analysis
PCs	Principal components
PET	Potential evapotranspiration
PI	Permeability Index
PLI	Pollution load index
PS	Potential salinity
PW	Piezometric well
RADAR	Radio detection and ranging
RBF	Radial basis function
RF	Random forest
RGB	Red, green, blue
RL	Reduced Level
RMSE	Root mean square error
RS	Remote sensing
RSBC	Residual sodium bicarbonate
RSC	Residual sodium carbonate
SAR	Sodium absorption ratio
SAR	Synthetic aperture radar
SARMA	Seasonal autoregressive moving average
SE	Standard error
SGWD	Stage of groundwater development
SNAP	Sentinel application platform
SO_4^{2-}	Sulfate
SoI	Survey of India
SPI	Standardized precipitation perspective
SPSS	Statistical package for the social sciences
SRTM	Shuttle radar topography mission
SSP	Soluble sodium percent
SSWSS	Spot source water supply scheme
SVM	Support vector machine
SVR	Systemic vascular resistance

SWID	State water investigation directorate
SWIR	Short-wave infrared imagery
Sy	Specific yield
TDS	Total dissolved solids
TH	Total hardness
TIRS	Thermal infrared sensor
TM	Thematic mapper
TOA	Top of atmosphere
TW	Tube well
UA	User accuracy
UNICEF	United Nations International Children's Emergency Fund
USGS	United States Geological Survey
USSL	United States Salinity Laboratory
UTM	Universal transverse mercator
VCI	Vegetation condition index
WGS84	The world geodetic system 1984
WHO	World health organization
WPI	Water poverty index
WQI	Water quality index
WTF	Water table fluctuation
XGB	Extreme gradient boosting

List of Figures

List of Tables

Chapter 1
Introduction

1.1 Backdrop

Water is vital for the survival of all organisms, including human beings, and is essential for maintaining environmental balance. The distribution, form, and availability of water resources are not uniform across the earth (Bhat, 2014). In the terrestrial system, the water resource is estimated as ~1.36×10^8 M ha-m of which 28% is occupied by freshwater and 0.6% of the freshwater is constituted by groundwater (Raghunath, 2006). Only 2.5% of the freshwater is accessible because a major portion of freshwater is stored in the form of glaciers and deep groundwater (Oki & Kanae, 2006). A sufficient amount and acceptable quality of water resources can greatly increase agricultural productivity while lowering the cost of treating water for industrial and drinking uses (Bilali & Taleb, 2020). The rapid growth of population, agricultural expansion, and increasing urbanization escalate the water demands which may accelerate the depletion of groundwater levels (GWL). Consequently, a drop in GWL causes major environmental risks, such as ecosystem degradation, low agricultural productivity, seawater intrusions, land subsidence, and declining groundwater quality (Li et al., 2019; Zhang et al., 2020; Sarkar et al., 2021). In addition, surface waterbodies provide freshwater for human and ecological systems (Huang et al., 2018). It has a paramount significance on regional climate, agriculture, and the environment (Yang et al., 2020). Similarly, it plays a fundamental role in the global carbon cycle and climate variations (Papa et al., 2022). Moreover, the areal extension of the surface waterbodies has become under threat due to the rapid growth of agricultural land, industrialization, and urbanization. Additionally, the quality of water resources is threatened due to natural and human activities. The physical, chemical, and microbiological characteristics of a given area determine the quality of the groundwater (Narasaiah & Rao, 2021). Geological elements, for instance, combined with human activities like farming, and the growth of the urban–industrial complex (Nas & Berktay, 2008; Subramani et al., 2005) dominate the

S. Mahammad, A. Islam, *Palaeochannels and Water Resource Management*,
https://doi.org/10.1007/978-3-031-92021-9_1

development and characteristics of groundwater. However, the quality of groundwater has an impact on society and human health (Milovanovic, 2007; Gidey, 2018).

The World Health Organization (WHO) states that the bulk (80%) of diseases that affect people worldwide is caused by contaminated water (Ganiyu et al., 2018). Furthermore, poor irrigation water deteriorates the soil's physical characteristics, which lowers agricultural output (Barik & Pattanayak, 2019). Furthermore, the growing use of fertilizers, pesticides, and waste from homes and businesses is degrading the quality of surface water. The high degree of climate unpredictability has a substantial impact on water resources, which results in social and economic instability in the local population's standard of living (Singh & Kumar, 2015). Floods and drought are common hydro-meteorological phenomena that occur due to abnormal behaviour of precipitation. Besides, anthropogenic activities near the fluvial system alter the morphological parameters of the river, causing congested flow of water and increasing siltation in the riverbeds, which may be responsible for flooding (Singh et al., 2020). Moreover, characteristics of geology, vegetation, and soil types and human activities such as overexploitation of water influence the propagation of drought conditions (West et al., 2019). Therefore, planning and monitoring of water resources are required to deal with the surplus and deficit conditions of the water. This requires the integration of principles of earth sciences, geography, environmental sciences, and geospatial sciences to comprehend the underlying principles governing the process and outcomes of the surface and groundwater utilizations at the global and regional scale in the wake of climate change and Anthropocene. Thus, this study is committed to uncovering these perspectives from an integrated outlook.

1.2 Previous Works

1.2.1 Worldwide Palaeochannels Study: A Bibliometric Analysis

Palaeochannels are widely discussed at the global scale with an escalation in the recent works as depicted by the SCOPUS datasets limited to English language documents (1958–2024) (Fig. 1.1a). In recent times, especially after 2000, there has been a great spurt in the palaeochannel research in the wake of climate extremes and anthropogenic interventions. Apart from this temporal rise, the spatial distribution of the work is also fascinating. The top three countries (the USA, Australia, and the UK) are followed by India, which demonstrates the significance of palaeochannel research in India (Fig. 1.1b). Regarding discipline, it is observed that earth and planetary sciences exhibit maximum works on palaeochannels followed by the environmental science disciplines (Fig. 1.1c). However, the field of palaeochannel research is quite diversified and maintains a close relationship with other areas of investigation which is vividly expressed in the word cloud (Fig. 1.2). Palaeochannel is in the centre with major links with aquifer, Holocene, groundwater, and

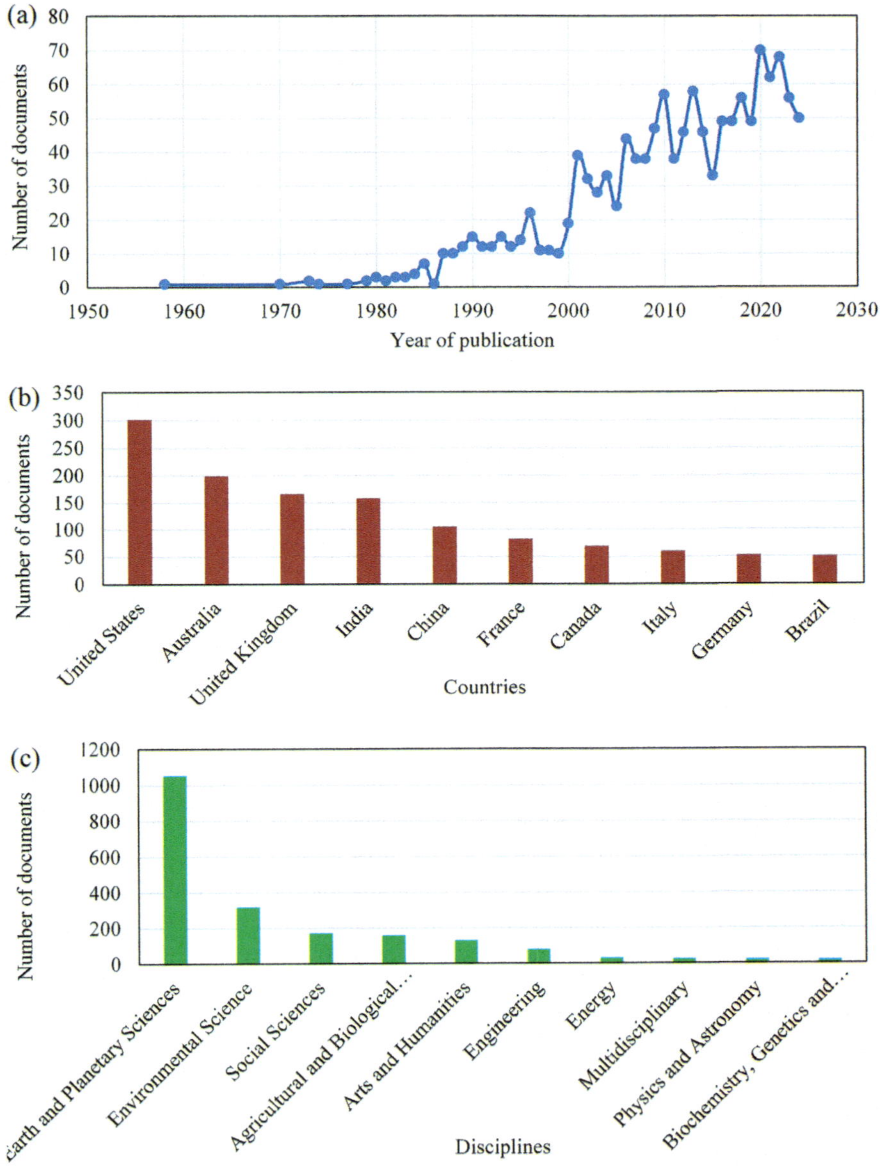

Fig. 1.1 Bibliometric analysis of the palaeochannel research at the global scale. (**a**) Trajectory of palaeochannel research from 1958 to 2024. (**b**) Top ten countries exhibiting palaeochannel research from 1958 to 2024. (**c**) Distribution of palaeochannel works in different disciplines

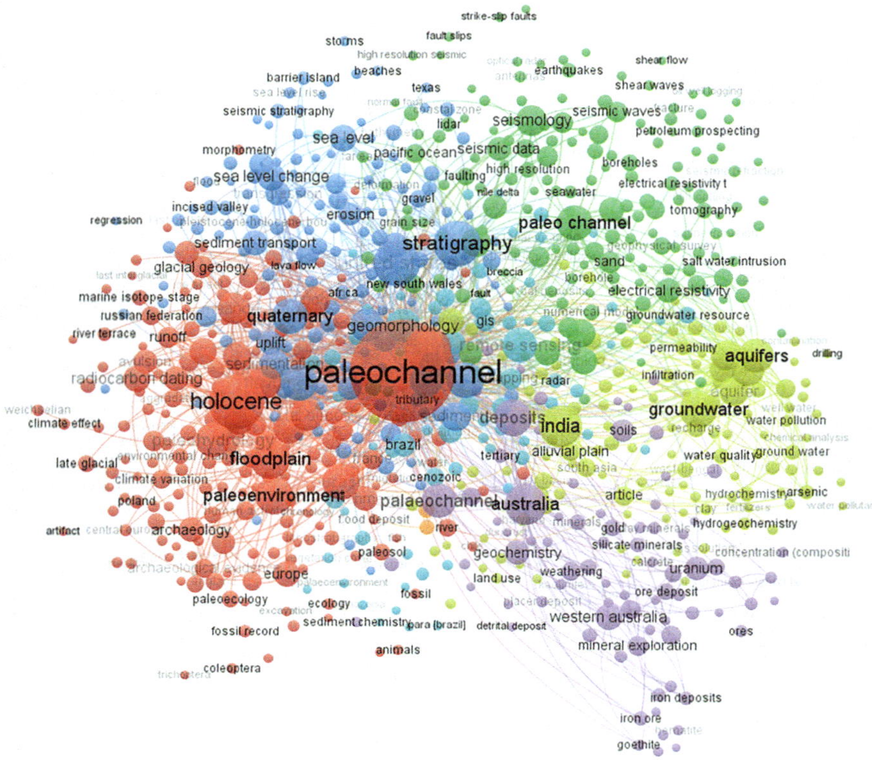

Fig. 1.2 Interdisciplinary nature of palaeochannels at the global scale. (Created using Vos-Viewer based on Scopus data (only English language documents) from 1958 to 2024

floodplains (Fig. 1.2). This indicates some crucial aspects of palaeochannel research floods and groundwater resource. However, climate extremes, floods, and droughts are intertwined to one another for our common future. There are very few studies that integrate palaeochannels and water resource management through the mitigation of floods and drought. This study is thus a novel attempt to integrate palaeochannel with water resource management.

1.2.2 *Identification of Palaeochannels*

Palaeochannels describe the remains of former channels that were filled in by newer alluvium, symbolizing the former fluvial system (Rathore et al., 2010; Rajawat et al., 2003). It has a significant impact on enhancing a region's groundwater supplies (Sinha et al., 2013) as well as offering palaeohydrological data (Kemp & Rhodes, 2010) and the places where archaeological sites are located (Roy & Sahu, 2016). Consequently, several scholars have made an effort to define palaeochannels using historical data in conjunction with GIS and geophysical methods. Globally, geophysical techniques were used in the majority of earlier investigations. While the geophysical techniques for palaeochannel detection are quite expensive and time-consuming, they yield precise results. With the development of contemporary technology, numerous researchers employed cutting-edge techniques, including radar with ground penetration (GPR) (Weymer et al., 2014), electrical resistivity sounding (Sinha et al., 2013), sub-surface stratigraphic records (Abbott, 2004), seismic refraction (Deen et al., 2000), electromagnetic induction (De Smedt et al. 2011), and sediment dates (Ejarque et al., 2014).

Remote sensing (RS) and geographic information systems (GIS) are vital for earth system monitoring (Sener et al., 2004). Many scholars have used the optical RS data to define palaeochannels based on the reflection characteristics of abandoned channels relative to the surrounding environment, along with a few image-processing approaches, including image categorization (Rajani & Rajawat, 2011), image transformation (Wray, 2009), the Normalized Difference Water Index (NDWI), Normalized Difference Vegetation Index (NDVI) (Roy & Sahu, 2016), contrast enhancement (Mehdi et al., 2016), Wang et al., 2011), principal component analysis (PCA) (Rajani & Rajawat, 2011), and decorrelation stretch (DCS) (Rathore et al., 2010).

Also, the stream network and hydrological routing have been created using the digital elevation model (DEM) (Mehdi et al., 2016; Elmahdy & Mohamed, 2015). High-resolution light detecting and ranging (LiDAR) data are popularly used for palaeochannel detection (Jones et al., 2007). Many researchers have also used radio detecting and ranging (RADAR) data of the C, X, and L bands in addition to multi-spectral satellite images (MSIs) to help in the precise identification of palaeochannel because these bands have more penetrating potential (Abdelkareem & El-Baz, 2014; Kumar & Rajawat, 2017).

1.2.3 Groundwater Dynamics and Quality Assessment

Physical descriptive models and data-driven statistical models are the two types of groundwater dynamics prediction models (Tian et al., 2016). The primary drawbacks of physical models include their propensity to require copious amounts of data on variables, including hydraulic conductivity, volumetric water content, groundwater extraction, precipitation, soil properties, and human activities like channel operations and dam construction (Zhou et al., 2020). In a region with little data and complex groundwater systems, data-driven statistical models are therefore regarded as an appealing substitute for groundwater depth estimates. To anticipate the GWL dynamics, a number of studies have employed traditional statistical models like multivariate linear regression models (MLR) in conjunction with time series models like the autoregressive integrated moving average (ARIMA) and seasonal autoregressive moving average (SARMA) models (Adamowski et al., 2012). ARIMA is a popular tool for social, technical, and economic problem prediction

(Saha et al., 2022), even though its accuracy and performance have not always been up to par (Yan & Ma, 2016). For GWL, the MLR has been used to forecast future events (Adamowski et al., 2012). ARIMA is frequently utilized in time series models for groundwater prediction and future projections for GWL, as well as in the prediction of social, engineering, and economic issues (Sahoo & Jha, 2013). Time series models offer an edge in prediction since they take into consideration correlations that may exist between data points. Various machine-learning techniques have been applied to forecast the dynamics of GWL as a result of the development of soft computational data-driven models in recent decades (Pham et al., 2022). In order to forecast GWL, Taormina et al. (2012) contrasted the feed-forward neural network (FFANN) with statistical models and regression models. Similar to this, Mohanty et al. (2015) used a neural network, while Li et al. (2016) used support vector machines (SVM) and artificial neural networks (ANN) to predict the GWL.

When compared to other water sources, groundwater is thought to be purer. Therefore, it satisfies the increasing population's need for drinking, agriculture, and industrial applications (Mahammad & Islam, 2021). As a result, its application to the socioeconomic advancement of populous nations like China and India is growing (Adimalla et al., 2018). Therefore, in many emerging nations, concerns about the quantity and quality of groundwater are now crucial to growth (Patel & Saha, 2022; Gidey, 2018). Numerous studies conducted worldwide have employed groundwater quality metrics to assess the suitability of groundwater for domestic, drinking, and irrigation purposes (Naidu et al., 2021; Salifu et al., 2015; Srivastava, 2019). These studies have also assessed the water's feasibility for irrigation through the application of diverse weathering indices. The water quality index (WQI), created by Horton (1965), is a widely used and important metric for assessing overall water quality (Tyagi et al., 2013). The weights of each physicochemical parameter for WQI are now assigned using the multi-criteria decision-making (MCDM) technique based on subjective (Bozdağ, 2014; Eldaw et al., 2021) and objective approaches (Hasan & Rai, 2020). In addition, a variety of irrigation hazard indices are frequently employed for groundwater irrigation suitability (Zhang et al., 2021; Paria et al., 2021; Zemunac et al., 2021). However, determining groundwater suitability based on discrete criteria frequently leads to incorrect categorization. Thus, in order to get rid of these issues, a comprehensive, parameter-based technique is developed to assess whether groundwater is appropriate for irrigation (Eldaw et al., 2021).

1.2.4 Surface Water Dynamics and Quality Assessment

Effectively identifying the geographical extent of surface waterbodies is vital for various scientific endeavours, including mapping surface water inventories, estimating drinking and irrigation water, mapping land use and land cover (LULC), detecting changes, and so on (Acharya et al., 2018). With its long-collected, reproducible, and even real-time data, remote sensing—a simpler, faster, and less expensive technique—offers significant capabilities for identifying, characterizing, and mapping surface waterbodies at local to global scales (Bijeesh & Narasimhamurthy, 2020). Surface waterbody spatial extension has been effectively detected using microwave remote sensing images (Goumehei et al., 2019). Because of their long wavelength radiation, microwave images may be seen through clouds and are usable in all types of weather, day or night. Because optical remote sensing images are more readily available and have appropriate spatial and temporal resolutions than microwave data, they are used far more frequently for surface waterbody detections (Huang et al., 2018). A multitude of studies have employed the single band-based technique of optical RS data to detect surface waterbodies (Feyisa et al., 2014), because, in comparison to multi-band approaches, it is more accurate and simpler to apply. Additionally, to find the surface waterbodies, spectral indices with a complex ratio of several bands have been used. McFeeters (1996) created the fundamental NDWI with Landsat TM images' green and near-infrared (NIR) bands. This index has been mostly utilized for extracting water features from the pictures. In a similar vein, other researchers have employed the Modified Normalized Difference Water Index (MNDWI), which was developed by Xu (2006), to locate surface waterbodies. Additionally, water characteristics have been extracted using Feyisa's et al. (2014) Automated Water Extraction Index. Additionally, surface water features have been found using NDVI which was introduced by Trucker and Sellers (1986).

Surface water is a vital component of the environment, necessary for life, and a significant supply of drinking and domestic water (Ustaoğlu et al., 2021). Both globally and regionally, its quality has declined. Because they are stagnant, standing water sources like lakes, ponds, etc. are more likely to become contaminated than moving water. Surface water pollution is caused by both natural and man-made processes (Kükrer & Mutlu, 2019). Moreover, by lowering oxygen levels and raising the eutrophication index, a rise in nutrients in surface water poses an ecological risk. For this reason, a large number of researchers have been hired to assess the surface water quality for irrigation and drinking. The Horton (1965) Water Quality Index, a commonly used technique, has been utilized to evaluate the drinking water quality using the Bureau of Indian Standards (BIS) and WHO's standard water quality indicators (Bouderbala, 2021; Panneerselvam et al. 2020; Wu et al., 2020; Aydin et al., 2020). A number of irrigation water quality indices, such as SAR, %Na, RSC, SSP, RSBC, MH, KR, PI, and PS, have been used to assess the irrigation water quality (Elsayed et al., 2020; Alsubih et al., 2022). Moreover, the Eutrophication Index (EI), Organic Pollution Index (OPI), and Overall Index of Pollution (OIP) have been used to evaluate the ecological risk of surface water (Sarkar and Islam 2020; Matta et al., 2018). Additionally, the background value of the heavy metal concentrations in terms of the dangers to human health has been taken into consideration when determining the heavy metal pollution index (Ali et al. 2019; Kumar et al. 2019).

1.2.5 Issues and Challenges of Water Services for Socio-Economic Development

The research work analysed the vital issues and challenges that resulted from the shortage of water resource and their relation to socio-economic development. Water is essential for socio-economic development. Katko and Hukka (2015) proposed a hierarchical framework that helps to obtain better water services and systems for socio-economic development. Vieira et al. (2017) analysed water use for different domestic purposes and compared their performance to quantify the potential saving of water resources. Moreover, sanitation and hygiene are vital indicators of human life. Mukherjee et al. (2020) efficiently measured the water security conditions related to water, sanitation, and hygiene in the slum area of Kolkata. Goswami and Ghosal (2022) studied that apart from the physical shortage of water, socio-economic capacity is responsible for causing water poverty in a region.

1.2.6 Palaeochannels in Water Resource Management

It is observed that palaeochannels are used to transfer water from the surplus area to the deficit area. The surplus area of water condition has been considered by the flood-prone region, whereas the deficit area has been determined by the drought-prone region of the study area. The potentiality of the palaeochannels has been assessed by the recharge rate of the underneath lithological layers based on the specific yield of materials and seasonal GWL fluctuation. Flood is the most devastating, frequent quasi-natural hazard resulting in highest number of deaths and property damage, especially in the low-lying floodplains and deltas (Sanyal & Lu, 2004). Therefore, measuring flood events has been done using different methods. The most reliable Gumbel and Log-Pearson type III methods have traditionally been used for flood frequency analysis based on long-term river discharge datasets (Islam & Sarkar, 2020; Das et al., 2020). In this method, data scarcity is the main constraint due to the lack of sufficient gauge stations. Apart from that, hydrodynamic modelling has been implemented in the lower Damodar Basin (Singh et al., 2020). Besides, RS and GIS have been used to detect flood inundation areas due to the time and cost-efficient nature of the remote sensing datasets. However, flood mainly occurs during the monsoon periods. Therefore, cloud cover causes serious constraints for flood analysis. To overcome such cumbersome, many of the studies implemented synthetic aperture radar (SAR) datasets for flood inundation mapping (Tiwari et al., 2020; Borah et al., 2018; Tripathi et al., 2020). Furthermore, drought conditions have been determined using several meteorological, agricultural, and hydrological indices (Son et al., 2021).

1.2.7 *Palaeochannels and Water Resource Management in the Damodar Fan Delta*

It is observed that many works have been conducted on the flood of the Damodar River (Ghosh, 2013; Singh et al., 2020; Ghosh & Guchhait, 2016; Bhattacharyya, 2011; Bera & Mistry, 2014). Ghosh and Mistri (2012a, b) identified channel and bank line shifting of the Damodar River. Similarly, several studies about the contamination of arsenic were conducted in the past (Acharyya & Shah, 2010; Pal & Mukherjee, 2010). Furthermore, the research area's agriculture has been adversely affected by the crucial development of groundwater (Majumder & Sivaramakrishnan, 2014). Studies on the palaeoenvironmental relevance of late quaternary deposits and the impact of palaeoclimatic influences on the fluvial structures deposited beneath the Damodar River‘s floodplain have been conducted (Ghosh & Guchhait, 2013). Moreover, Mallick and Niyogi (1972) and Niyogi (1975) examined quaternary geology and supplied the DFD‘s geomorphological units and palaeochannel distribution. Moreover, Ghosh and Jana (2019) investigated how historical data affected the DFD‘s palaeochannel delineation. Based on previously published works, they have pinpointed the historical locations referenced in the folktale from Bengal to illustrate the linear course of the ancient Damodar River.

1.3 Research Gaps and Study Aims

It seems that the research area's palaeochannels were defined using photo-geomorphic and geological analysis; however, geospatial-based low-cost detection techniques within a short time are rare in the context of DFD. Moreover, the dynamics of surface and groundwater quantity and quality in different seasons and their impacts on society and the environment are unknown. Another important research gap shows that there are no such previous attempts to relate groundwater and surface water dynamics with the existence of palaeochannels. Previous studies indicated the relationship between groundwater arsenic contamination and palaeochannel locations; however, no such studies have been found to concentrate on groundwater and surface water quality assessment for irrigation, drinking, and ecological purposes in relation to the locations of the palaeochannels. While the study region is aware of the hydrological extremes, such as floods and the rapidly declining groundwater level, appropriate use of the palaeochannels already in place to lessen the severity of floods and droughts has not been tried as of yet. In other words, no previous framework has been observed, which can integrate groundwater and surface water dynamics in the context of distributing excess monsoon floodwater into water-stressed areas of the DFD.

This work focuses on the identification and utilization of palaeochannels for water resource management. Therefore, the major objectives of this study can be pointed out as follows:

(a) To identify the palaeochannels using historical records, geospatial, and geological evidence.
(b) To assess the dynamics of groundwater and its quality for drinking, and irrigation purposes.
(c) To evaluate the dynamics of surface water and its quality for drinking, and irrigation purposes.
(d) To portray the potentiality of palaeochannels in water resource management.

1.4 The Study Area

1.4.1 Locational Attributes

The Damodar River formed two alluvial fans: (1) the Memari fan trending east and (2) the Tarakeswar fan trending towards the south (Acharyya & Shah, 2006; Mallick & Niyogi, 1972; Niyogi, 1975). The Damodar Fan Delta is selected as the study area is located in Eastern India (Fig. 1.3). It is situated in the eastern Hooghly River interfluve and the western Damodar River interfluve. The Kusumgram fan encircles it from the north. The Holocene deposit distinguishes the younger delta plain of the Damodar Fan Delta.

The topographic configuration of the DFD is really fascinating from the perspective of longitudinal and cross-sectional dimensions. To this end, relief profiles across the DFD have been drawn along the lines A–A1 and B–B1 (Fig. 1.1). The longitudinal profile (A–A1) extending from the DFD‘s apex part to its toe area is nearly concave due to extensive erosion caused by the Ajay River at the northern boundary and the Dwarkeswar River at the southern boundary. In contrast, the cross profile (B–B1) extending from the DFD’s northern part to its southern boundary is perfectly convex (Mahata & Maiti, 2019).

1.4.2 Justification for Selection

The Damodar Fan Delta is selected as the study area based on the following criteria:

(a) The Damodar River along with its distributaries changes its courses through the centuries and leaves behind numerous palaeochannels in the Damodar Fan Delta.
(b) Several floods occurred in the study area throughout the centuries.
(c) The water quality of the study area is associated with arsenic contamination.

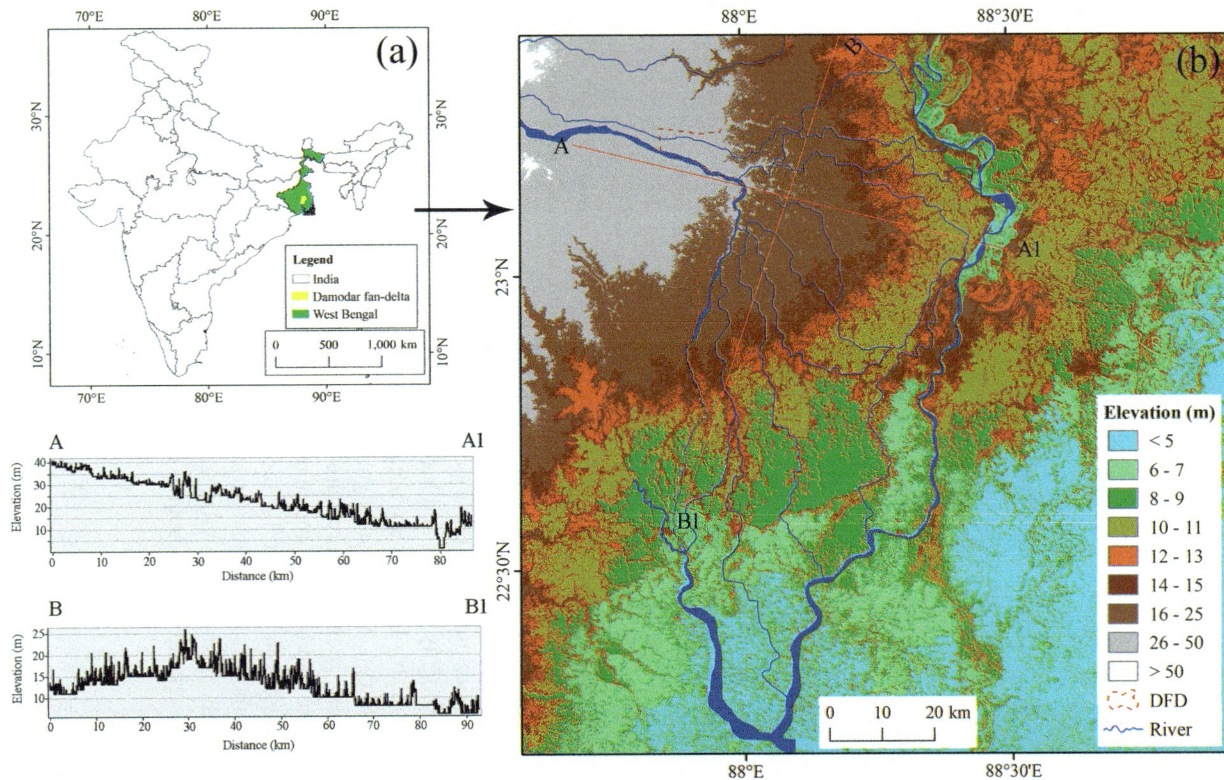

Fig. 1.3 Location map of the study area, (a) Location of study area within India, (b) Boundary and relief of the Damodar Fan Delta (Source: Elevations are extracted from SRTM DEM, the boundary of Damodar Fan Delta is modified from Mallick and Niyogi (1972) and Acharyya and Shah (2006))

(d) The 13 semi-critical blocks of Purba Barddhaman and Hooghly districts classified by the central groundwater board are located in the study area.
(e) The agro-based area requires huge amounts of irrigation water, especially in the lean months.

1.5 The Need for This Work

This research deals with the identification of palaeochannels and suggests their optimum uses in water resources management. The study area is located in the lower Damodar River Basin. The Damodar River is known for its devastating flood with a different magnitude from the historical past. Therefore, it was named Sorrow of Bengal (Rudra, 2008). The lower Damodar basin is characterized by higher fertility and agricultural potentiality which make the region famous for human settlement and agricultural activities (Ghosh & Guchhait, 2016). Therefore, the flood condition tremendously affects the society and economy of the region. Moreover, the region has been experiencing a gradual depletion of GWL for the last few decades. The Central Ground Water Board (CGWB) revealed that the number of semi-critical community development (CD) blocks in the research area had significantly increased due to groundwater depletion. According to the CGWB, in 2004 there were only two semi-critical CD blocks in the research area, but by 2013, there were 13 (CGWB, 2006, 2017). Besides, the Damodar Fan Delta (DFD) is facing serious constraints related to water quality, especially arsenic contamination (Pal & Mukherjee, 2010; Acharyya & Shah, 2006). In addition, this study area experienced numerous channel shifting in the Damodar River, mainly its lower courses (Ghosh & Mistri, 2012a, b). Therefore, changing the courses of the Damodar River along with its tributaries modifies the topographic configuration of the Barddhaman, Howrah, and Hooghly districts and leaves behind numerous palaeochannels (Ghosh & Mistri, 2012a, b). In the study area, the palaeochannels both buried and abandoned channels can be used as significant pathways to pass the floodwater and potential artificial recharge site to prevent the depletion of GWL. Thus, this study will, on the one hand, reduce the magnitude of flood dimension and on the other increase the irrigation potential for the ago-based Damodar Fan Delta by maintaining a good balance of surface and groundwater.

Therefore, the next chapter will sketch out the study area profile in the context of the palaeochannel and water resource management.

References

Abbott, S. (2004). Airborne geophysical data, including electromagnetics, enhances rapid catchment appraisal in the upper kent recovery catchment. *Regolith*, 2–6.

Abdelkareem, M., & El-Baz, F. (2014). Analyses of optical images and radar data reveal structural features and predict groundwater accumulations in the central Eastern Desert of Egypt. *Arabian Journal of Geosciences, 8*, 2653–2666. https://doi.org/10.1007/s12517-014-1434-7

Acharya, T., Subedi, A., & Lee, D. (2018). Evaluation of water indices for surface water extraction in a Landsat 8 scene of Nepal. *Sensors, 18*, 2580. https://doi.org/10.3390/s18082580

Acharyya, S. K., & Shah, B. A. (2006). Arsenic-contaminated groundwater from parts of Damodar fan-delta and west of Bhagirathi River, West Bengal, India: Influence of fluvial geomorphology and quaternary morphostratigraphy. *Environmental Geology, 52*, 489–501. https://doi.org/10.1007/s00254-006-0482-z

Acharyya, S. K., & Shah, B. A. (2010). Groundwater arsenic pollution affecting deltaic West Bengal, India. *Current Science (Bangalore), 99*, 1787–1794.

Adamowski, J., Chan, H. F., Prasher, S. O., Ozga-Zielinski, B., & Sliusarieva, A. (2012). Comparison of multiple linear and nonlinear regression, autoregressive integrated moving average, artificial neural network, and wavelet artificial neural network methods for urban water demand forecasting in Montreal, Canada. *Water Resources Research, 48*. https://doi.org/10.1029/2010wr009945

Adimalla, N., Li, P., & Venkatayogi, S. (2018). Hydrogeochemical evaluation of groundwater quality for drinking and irrigation purposes and integrated interpretation with water quality index studies. *Environmental Processes, 5*, 363–383. https://doi.org/10.1007/s40710-018-0297-4

Ali, M. M., Ali, M. L., Proshad, R., Islam, S., Rahman, Z., Tusher, T. R., … Al, M. A. (2019). Heavy metal concentrations in commercially valuable fishes with health hazard inference from Karnaphuli river, Bangladesh. *Human and Ecological Risk Assessment: An International Journal, 26*(10), 2646–2662. https://doi.org/10.1080/10807039.2019.1676635

Alsubih, M., Mallick, J., Islam, A. R., Almesfer, M. K., Kahla, N. B., Talukdar, S., & Ahmed, M. (2022). Assessing surface water quality for irrigation purposes in some dams of Asir Region, Saudi Arabia using multi-statistical modeling approaches. *Water, 14*, 1439. https://doi.org/10.3390/w14091439

Aydin, H., Ustaoğlu, F., Tepe, Y., & Soylu, E. N. (2020). Assessment of water quality of streams in Northeast Turkey by water quality index and multiple statistical methods. *Environmental Forensics, 22*, 270–287. https://doi.org/10.1080/15275922.2020.1836074

Barik, R., & Pattanayak, S. K. (2019). Assessment of groundwater quality for irrigation of green spaces in the Rourkela city of Odisha, India. *Groundwater for Sustainable Development, 8*, 428–438. https://doi.org/10.1016/j.gsd.2019.01.005

Bera, S., & Mistry, B. (2014). Flood in the lower Damodar Basin and channel morphology: A case study at the bifurcation zone into Damodar and Mundeswari river, West Bengal. *International Journal of Geology, Earth & Environmental Sciences, 4*, 172–181.

Bhat, T. A. (2014). An analysis of demand and supply of water in India. *Journal of Environment and Earth Science, 4*(11), 67–72.

Bhattacharyya, K. (2011). *The Lower Damodar River, India: Understanding the human role in changing fluvial environment*. Springer.

Bijeesh, T. V., & Narasimhamurthy, K. N. (2020). Surface water detection and delineation using remote sensing images: A review of methods and algorithms. *Sustainable Water Resources Management, 6*. https://doi.org/10.1007/s40899-020-00425-4

Bilali, A. E., & Taleb, A. (2020). Prediction of irrigation water quality parameters using machine learning models in a semi-arid environment. *Journal of the Saudi Society of Agricultural Sciences, 19*, 439–451. https://doi.org/10.1016/j.jssas.2020.08.001

Borah, S. B., Sivasankar, T., Ramya, M. N., & Raju, P. L. (2018). Flood inundation mapping and monitoring in Kaziranga National Park, Assam using Sentinel-1 SAR data. *Environmental Monitoring and Assessment, 190*. https://doi.org/10.1007/s10661-018-6893-y

Bouderbala, A. (2021). Index methods for the assessment of surface water quality: The case study of Oued Fodda dam, in the Northwest of Algeria. *Environment, Development and Sustainability, 23*, 13340–13363. https://doi.org/10.1007/s10668-020-01215-w

Bozdağ, A. (2014). Combining AHP with GIS for assessment of irrigation water quality in Çumra irrigation district (Konya), Central Anatolia, Turkey. *Environmental Earth Sciences, 73*, 8217–8236. https://doi.org/10.1007/s12665-014-3972-4

CGWB. (2006). *Dynamic ground water resources of India.* Ministry of Water Resources, River Development & Ganga Rejuvenation, Government of India, Faridabad.

CGWB. (2017). *Dynamic ground water resources of India.* Ministry of Water Resources, River Development & Ganga Rejuvenation, Government of India, Faridabad.

Das, J., Rahman, A. T., Mandal, T., & Saha, P. (2020). Exploring driving forces of large-scale unsustainable groundwater development for irrigation in lower Ganga River basin in India. *Environment, Development and Sustainability, 23*, 7289–7309. https://doi.org/10.1007/s10668-020-00917-5

Deen, T. J., Gohl, K., Leslie, C., Papp, E., & Wake-Dyster, K. (2000). Seismic refraction inversion of a palaeochannel system in the Lachlan Fold Belt, Central New South Wales. *Exploration Geophysics, 31*, 389–393. https://doi.org/10.1071/eg00389

De Smedt, P., Van Meirvenne, M., Meerschman, E., Saey, T., Bats, M., Court-Picon, M., De Reu, J., Zwertvaegher, A., Antrop, M., Bourgeois, J., De Maeyer, P., Finke, P. A., Verniers, J., & Crombé, P. (2011). Reconstructing palaeochannel morphology with a mobile multicoil electromagnetic induction sensor. *Geomorphology, 130*(3-4), 136–141. https://doi.org/10.1016/j.geomorph.2011.03.009

Ejarque, A., Beauger, A., Miras, Y., Peiry, J.-L., Voldoire, O., Vautier, F., et al. (2014). Historical fluvial palaeodynamics and multi-proxy palaeoenvironmental analyses of a palaeochannel, Allier River, France. *Geodinamica Acta, 27*, 25–47. https://doi.org/10.1080/09853111.2013.877232

Eldaw, E., Huang, T., Mohamed, A. K., & Mahama, Y. (2021). Classification of groundwater suitability for irrigation purposes using a comprehensive approach based on the AHP and GIS techniques in North Kurdufan Province, Sudan. *Applied Water Science, 11*. https://doi.org/10.1007/s13201-021-01443-z

Elmahdy, S. I., & Mohamed, M. M. (2015). Remote sensing and geophysical survey applications for delineating near-surface palaeochannels and shallow aquifer in The United Arab Emirates. *Geocarto International, 30*, 723–736. https://doi.org/10.1080/10106049.2014.997306

Elsayed, S., Hussein, H., Moghanm, F. S., Khedher, K. M., Eid, E. M., & Gad, M. (2020). Application of irrigation water quality indices and multivariate statistical techniques for surface water quality assessments in the northern Nile Delta, Egypt. *Water, 12*, 3300. https://doi.org/10.3390/w12123300

Feyisa, G. L., Meilby, H., Fensholt, R., & Proud, S. R. (2014). Automated water extraction index: A new technique for surface water mapping using Landsat imagery. *Remote Sensing of Environment, 140*, 23–35. https://doi.org/10.1016/j.rse.2013.08.029

Ganiyu, S. A., Badmus, B. S., Olurin, O. T., & Ojekunle, Z. O. (2018). Evaluation of seasonal variation of water quality using multivariate statistical analysis and irrigation parameter indices in Ajakanga area, Ibadan, Nigeria. *Applied Water Science, 8*. https://doi.org/10.1007/s13201-018-0677-y

Ghosh, S. (2013). Estimation of flash flood magnitude and flood risk in the lower segment of Damodar River basin, India. *International Journal of Geology, Earth & Environmental Science, 3*, 97–114.

Ghosh, S., & Guchhait, S. K. (2013). Palaeoenvironmental significance of fluvial facies and archives of late quaternary deposits in the floodplain of Damodar River, India. *Arabian Journal of Geosciences, 7*, 4145–4161. https://doi.org/10.1007/s12517-013-1079-y

Ghosh, S., & Guchhait, S. K. (2016). Dam-induced changes in flood hydrology and flood frequency of tropical river: A study in Damodar River of West Bengal, India. *Arabian Journal of Geosciences, 9*, 1–26.

Ghosh, P. K., & Jana, N. C. (2019). Historical evidences in the identification of palaeochannels of Damodar River in Western Ganga-Brahmaputra Delta. In *Geography of the physical environment* (pp. 127–138). Springer International Publishing. https://doi.org/10.1007/978-3-319-90427-6_7

Ghosh, S., & Mistri, B. (2012a). Hydrogeomorphic significance of sinuosity index in relation to river instability: A case study of Damodar River, West Bengal, India. *International Journal of Advances in Earth Sciences, 1*, 49–57.

Ghosh, S., & Mistri, B. (2012b). Reconstructing the phases of channel shifting through identification of palaeochannels and historical accounts of extreme floods of Damodar River in West Bengal. *Indian Journal of Geomorphology, 17*, 65–80.

Gidey, A. (2018). Geospatial distribution modeling and determining suitability of groundwater quality for irrigation purpose using geospatial methods and water quality index (WQI) in Northern Ethiopia. *Applied Water Science, 8*. https://doi.org/10.1007/s13201-018-0722-x

Goswami, T., & Ghosal, S. (2022). Domestic water poverty in a semi-arid district of eastern India: Multiple dimensions, regional pattern, and association with human development. *Environmental Development, 44*, 100742. https://doi.org/10.1016/j.envdev.2022.100742

Goumehei, E., Tolpekin, V., Stein, A., & Yan, W. (2019). Surface water body detection in polarimetric SAR data using contextual complex Wishart classification. *Water Resources Research, 55*, 7047–7059. https://doi.org/10.1029/2019wr025192

Hasan, M. S., & Rai, A. K. (2020). Groundwater quality assessment in the lower Ganga Basin using entropy information theory and GIS. *Journal of Cleaner Production, 274*, 123077. https://doi.org/10.1016/j.jclepro.2020.123077

Horton, R. K. (1965). An index number system for rating water quality. *Journal - Water Pollution Control Federation, 37*, 300–306.

Huang, W., DeVries, B., Huang, C., Lang, M., Jones, J., Creed, I., & Carroll, M. (2018). Automated extraction of surface water extent from Sentinel-1 data. *Remote Sensing, 10*, 797. https://doi.org/10.3390/rs10050797

Islam, A., & Sarkar, B. (2020). Analysing flood history and simulating the nature of future floods using Gumbel method and log-Pearson type III: The case of the Mayurakshi River basin, India. *Bulletin of Geography. Physical Geography Series, 19*, 43–69.

Jones, A. F., Brewer, P. A., Johnstone, E., & Macklin, M. G. (2007). High-resolution interpretative geomorphological mapping of river valley environments using airborne LiDAR data. *Earth Surface Processes and Landforms, 32*, 1574–1592. https://doi.org/10.1002/esp.1505

Katko, T. S., & Hukka, J. J. (2015). Social and economic importance of water services in the built environment: Need for more structured thinking. *Procedia Economics and Finance, 21*, 217–223. https://doi.org/10.1016/s2212-5671(15)00170-7

Kemp, J., & Rhodes, E. J. (2010). Episodic fluvial activity of inland rivers in southeastern Australia: Palaeochannel systems and terraces of the Lachlan River. *Quaternary Science Reviews, 29*, 732–752. https://doi.org/10.1016/j.quascirev.2009.12.001

Kükrer, S., & Mutlu, E. (2019). Assessment of surface water quality using water quality index and multivariate statistical analyses in Saraydüzü dam Lake, Turkey. *Environmental Monitoring and Assessment, 191*. https://doi.org/10.1007/s10661-019-7197-6

Kumar, H., & Rajawat, A. S. (2017). Potential of RISAT-1 SAR data in detecting palaeochannels in parts of the Thar Desert, India. *Current Science, 113*, 1899. https://doi.org/10.18520/cs/v113/i10/1899-1905

Kumar, V., Parihar, R. D., Sharma, A., Bakshi, P., Singh Sidhu, G. P., Bali, A. S., Karaouzas, I., Bhardwaj, R., Thukral, A. K., Gyasi-Agyei, Y., & Rodrigo-Comino, J. (2019). Global evaluation of heavy metal content in surface water bodies: A metaanalysis using heavy metal pollution indices and multivariate statistical analyses. *Chemosphere, 236*, 124364. https://doi.org/10.1016/j.chemosphere.2019.124364

Li, B., Yang, G., Wan, R., Dai, X., & Zhang, Y. (2016). Comparison of random forests and other statistical methods for the prediction of Lake water level: A case study of the Poyang lake in China. *Hydrology Research, 47*, 69–83. https://doi.org/10.2166/nh.2016.264

Li, L., Zheng, Y., & Li. (2019). Groundwater level prediction for the arid oasis of Northwest China based on the artificial bee Colony algorithm and a Back-propagation neural network with double hidden layers. *Water, 11*, 860. https://doi.org/10.3390/w11040860

Mahammad, S., & Islam, A. (2021). Evaluating the groundwater quality of Damodar Fan Delta (India) using fuzzy-AHP MCDM technique. *Applied Water Science, 11*. https://doi.org/10.1007/s13201-021-01408-2

Mahata, H. K., & Maiti, R. (2019). Evolution of Damodar Fan Delta in the Western Bengal Basin, West Bengal. *Journal of the Geological Society of India, 93*, 645–656.

Majumder, A., & Sivaramakrishnan, L. (2014). Ground water budgeting in alluvial Damodar fan delta: A study in semi-critical Pandua block of West Bengal, India. *International Journal of Geology, Earth & Environmental Sciences, 4*, 23–37.

Mallick, S., & Niyogi, D. (1972). Application of geomorphology in groundwater prospecting in the alluvial plains around Burdwan, West Bengal. *Indian Geohydrology, 8*(2), 86–98.

Matta, G., Naik, P. K., Machell, J., Kumar, A., Gjyli, L., Tiwari, A. K., & Kumar, A. (2018). Comparative study on seasonal variation in hydro-chemical parameters of Ganga River water using comprehensive pollution index (CPI) at Rishikesh, (Uttarakhand) India. *Desalination and Water Treatment, 118*, 87–95. https://doi.org/10.5004/dwt.2018.22487

McFeeters, S. K. (1996). The use of the normalized difference water index (NDWI) in the delineation of open water features. *International Journal of Remote Sensing, 17*, 1425–1432. https://doi.org/10.1080/01431169608948714

Mehdi, S. M., Pant, N. C., Saini, H. S., Mujtaba, S. A., & Pande, P. (2016). Identification of palaeochannel configuration in the Saraswati River basin in parts of Haryana and Rajasthan, India, through digital remote sensing and GIS. *Episodes Journal of International Geoscience, 39*, 29–38.

Milovanovic, M. (2007). Water quality assessment and determination of pollution sources along the Axios/Vardar River, southeastern Europe. *Desalination, 213*, 159–173. https://doi.org/10.1016/j.desal.2006.06.022

Mohanty, S., Jha, M. K., Raul, S. K., Panda, R. K., & Sudheer, K. P. (2015). Using artificial neural network approach for simultaneous forecasting of weekly groundwater levels at multiple sites. *Water Resources Management, 29*, 5521–5532. https://doi.org/10.1007/s11269-015-1132-6

Mukherjee, S., Sundberg, T., & Schütt, B. (2020). Assessment of water security in socially excluded areas in Kolkata, India: An approach focusing on water, sanitation and hygiene. *Water, 12*, 746. https://doi.org/10.3390/w12030746

Naidu, S., Gupta, G., Singh, R., Tahama, K., & Erram, V. C. (2021). Hydrogeochemical processes regulating the groundwater quality and its suitability for drinking and irrigation purpose in parts of coastal Sindhudurg District, Maharashtra. *Journal of the Geological Society of India, 97*, 173–185. https://doi.org/10.1007/s12594-021-1649-7

Narasaiah, V., & Rao, B. V. (2021). Groundwater quality of an hard rock aquifer in the Subledu Basin of Khammam district, India. *Applied Water Science, 11*. https://doi.org/10.1007/s13201-021-01424-2

Nas, B., & Berktay, A. (2008). Groundwater quality mapping in urban groundwater using GIS. *Environmental Monitoring and Assessment, 160*, 215–227. https://doi.org/10.1007/s10661-008-0689-4

Niyogi, D. (1975). Quaternary geology of the coastal plain of West Bengal and Orissa. *Indian Journal of Earth Science, 2*, 51–61.

Oki, T., & Kanae, S. (2006). Global hydrological cycles and world water resources. *Science, 313*(5790), 1068–1072.

Pal, T., & Mukherjee, P. K. (2010). Search for groundwater arsenic in Pleistocene sequence of the Damodar River flood plain, West Bengal. *Indian Journal Of Geosciences, 64*, 109–112.

Panneerselvam. B., Karuppannan. S., Muniraj. K. (2020). Evaluation of drinking and irrigation suitability of groundwater with special emphasizing the health risk posed by nitrate contamination using nitrate pollution index (NPI) and human health risk assessment (HHRA). *Human and Ecological Risk Assessment: An International Journal, 27(5)*, 1324–1348. https://doi.org/10.1080/10807039.2020.183330010

Papa, F., Crétaux, J.-F., Grippa, M., Robert, E., Trigg, M., Tshimanga, R. M., et al. (2022). Water resources in Africa under global change: Monitoring surface waters from space. *Surveys in Geophysics*. https://doi.org/10.1007/s10712-022-09700-9

Paria, B., Pani, A., Mishra, P., & Behera, B. (2021). Irrigation-based agricultural intensification and future groundwater potentiality: Experience of Indian states. *SN Applied Sciences, 3*. https://doi.org/10.1007/s42452-021-04417-7

Patel, P. M., & Saha, D. (2022). Groundwater: A juggernaut of socio-economic development and stability in the arid region of Kachchh. In *Groundwater for sustainable livelihoods and equitable growth* (pp. 231–252). CRC Press.

Pham, Q. B., Kumar, M., Nunno, F. D., Elbeltagi, A., Granata, F., Islam, A. R., et al. (2022). Groundwater level prediction using machine learning algorithms in a drought-prone area. *Neural Computing and Applications, 34*, 10751–10773. https://doi.org/10.1007/s00521-022-07009-7

Raghunath, H. M. (2006). *Hydrology: Principles, analysis and design*. New Age International.

Rajani, M. B., & Rajawat, A. S. (2011). Potential of satellite based sensors for studying distribution of archaeological sites along palaeo channels: Harappan sites a case study. *Journal of Archaeological Science, 38*, 2010–2016. https://doi.org/10.1016/j.jas.2010.08.008

Rajawat, A. S., Verma, P. K., Nayak, S., Rajawat, A. S., Verma, P. K., & Nayak, S. (2003). Reconstruction of palaeodrainage network in Northwest India: Retrospect and prospects of remote sensing based studies. *Proceedings-Indian National Science Academy Part A, 69*, 217–230.

Rathore, V. S., Nathawat, M. S., & Champatiray, P. K. (2010). Palaeochannel detection and aquifer performance assessment in Mendha River catchment, Western India. *Journal of Hydrology, 395*, 216–225. https://doi.org/10.1016/j.jhydrol.2010.10.026

Roy, S., & Sahu, A. S. (2016). Palaeo-path investigation of the lower Ajay River (India) using archaeological evidence and applied remote sensing. *Geocarto International, 31*, 966–984. https://doi.org/10.1080/10106049.2015.1094526

Rudra, K. (2008). Banglar Nadikatha (in Bengali). *Banglar Nadikatha (in Bengali)*. Sahitya Samsad Kolkata.

Saha, U. D., Saheb, A. M., Islam, A., Barman, S. D., Dutt, S., & Islam, A. R. (2022). Characterizing the trend of channel braiding of a tropical transboundary river using spatial growth component analysis and ARIMA model. *Advances in Space Research, 70*, 1773–1794. https://doi.org/10.1016/j.asr.2022.06.032

Sahoo, S., & Jha, M. K. (2013). Groundwater-level prediction using multiple linear regression and artificial neural network techniques: A comparative assessment. *Hydrogeology Journal, 21*, 1865–1887. https://doi.org/10.1007/s10040-013-1029-5

Salifu, M., Aidoo, F., Hayford, M. S., Adomako, D., & Asare, E. (2015). Evaluating the suitability of groundwater for irrigational purposes in some selected districts of the upper West region of Ghana. *Applied Water Science, 7*, 653–662. https://doi.org/10.1007/s13201-015-0277-z

Sanyal, J., & Lu, X. X. (2004). Application of remote sensing in flood management with special reference to monsoon Asia: A review. *Natural Hazards, 33*, 283–301. https://doi.org/10.1023/b:nhaz.0000037035.65105.95

Sarkar, B., Islam, A. (2020). Drivers of water pollution and evaluating its ecological stress with special reference to macrovertebrates (fish community structure): a case of Churni River, India. *Environ Monit Assess, 192*, 45. https://doi.org/10.1007/s10661-019-7988-9

Sarkar, B., Islam, A., & Majumder, A. (2021). Seawater intrusion into groundwater and its impact on irrigation and agriculture: Evidence from the coastal region of West Bengal, India. *Regional Studies in Marine Science, 44*, 101751. https://doi.org/10.1016/j.rsma.2021.101751

Sener, E., Davraz, A., & Ozcelik, M. (2004). An integration of GIS and remote sensing in groundwater investigations: A case study in Burdur, Turkey. *Hydrogeology Journal, 13*, 826–834. https://doi.org/10.1007/s10040-004-0378-5

Singh, R. B., & Kumar, A. (2015). Climate variability and water resource scarcity in drylands of Rajasthan, India. *Geoenvironmental Disasters, 2*. https://doi.org/10.1186/s40677-015-0018-5

Singh, R. K., Villuri, V. G., Pasupuleti, S., & Nune, R. (2020). Hydrodynamic modeling for identifying flood vulnerability zones in lower Damodar river of eastern India. *Ain Shams Engineering Journal, 11*, 1035–1046. https://doi.org/10.1016/j.asej.2020.01.011

Sinha, R., Yadav, G. S., Gupta, S., Singh, A., & Lahiri, S. K. (2013). Geo-electric resistivity evidence for subsurface palaeochannel systems adjacent to Harappan sites in Northwest India. *Quaternary International, 308-309*, 66–75. https://doi.org/10.1016/j.quaint.2012.08.002

Son, B., Park, S., Im, J., Park, S., Ke, Y., & Quackenbush, L. J. (2021). A new drought monitoring approach: Vector projection analysis (VPA). *Remote Sensing of Environment, 252*, 112145. https://doi.org/10.1016/j.rse.2020.112145

Srivastava, S. K. (2019). Assessment of groundwater quality for the suitability of irrigation and its impacts on crop yields in the Guna district, India. *Agricultural Water Management, 216*, 224–241. https://doi.org/10.1016/j.agwat.2019.02.005

Subramani, T., Elango, L., & Damodarasamy, S. R. (2005). Groundwater quality and its suitability for drinking and agricultural use in Chithar River Basin, Tamil Nadu, India. *Environmental Geology, 47*, 1099–1110. https://doi.org/10.1007/s00254-005-1243-0

Taormina, R., Chau, K.-W., & Sethi, R. (2012). Artificial neural network simulation of hourly groundwater levels in a coastal aquifer system of the Venice lagoon. *Engineering Applications of Artificial Intelligence, 25*, 1670–1676. https://doi.org/10.1016/j.engappai.2012.02.009

Tian, J., Li, C., Liu, J., Yu, F., Cheng, S., Zhao, N., & Jaafar, W. W. (2016). Groundwater depth prediction using data-driven models with the assistance of gamma test. *Sustainability, 8*, 1076. https://doi.org/10.3390/su8111076

Tiwari, V., Kumar, V., Matin, M. A., Thapa, A., Ellenburg, W. L., Gupta, N., & Thapa, S. (2020). Flood inundation mapping-Kerala 2018; harnessing the power of SAR, automatic threshold detection method and Google Earth Engine . (G. J.-P. Schumann, Ed.). *PLoS One, 15*, e0237324. https://doi.org/10.1371/journal.pone.0237324

Tripathi, G., Pandey, A. C., Parida, B. R., & Kumar, A. (2020). Flood inundation mapping and impact assessment using multi-temporal optical and SAR satellite data: A case study of 2017 flood in Darbhanga District, Bihar, India. *Water Resources Management, 34*, 1871–1892. https://doi.org/10.1007/s11269-020-02534-3

Trucker, C. J., & Sellers, P. J. (1986). Satellite remote sensing of primary production. *International Journal of Remote Sensing, 7*, 1395–1416. https://doi.org/10.1080/01431168608948944

Tyagi, S., Sharma, B., Singh, P., & Dobhal, R. (2013). Water quality assessment in terms of water quality index. *American Journal of water resources, 1*, 34–38.

Ustaoğlu, F., Taş, B., Tepe, Y., & Topaldemir, H. (2021). Comprehensive assessment of water quality and associated health risk by using physicochemical quality indices and multivariate analysis in Terme River, Turkey. *Environmental Science and Pollution Research, 28*, 62736–62754. https://doi.org/10.1007/s11356-021-15135-3

Vieira, P., Jorge, C., & Covas, D. (2017). Assessment of household water use efficiency using performance indices. *Resources, Conservation and Recycling, 116*, 94–106. https://doi.org/10.1016/j.resconrec.2016.09.007

Wang, X., Guo, Z., Wu, L., Zhu, C., & He, H. (2011). Extraction of palaeochannel information from remote sensing imagery in the east of Chaohu Lake, China. *Frontiers of Earth Science, 6*, 75–82. https://doi.org/10.1007/s11707-011-0188-8

West, H., Quinn, N., & Horswell, M. (2019). Remote sensing for drought monitoring & impact assessment: Progress, past challenges and future opportunities. *Remote Sensing of Environment, 232*, 111291. https://doi.org/10.1016/j.rse.2019.111291

Weymer, B., Houser, C., Giardino, J., Barrineau, P., Everett, M., & Bishop, M. (2014). The role of geologic inheritance on storm impacts along the south Texas Coast, USA. *EGU General Assembly Conference Abstracts*, (p. 108).

Wray, R. A. (2009). Palaeochannels of the Namoi River floodplain, New South Wales, Australia: The use of multispectral Landsat imagery to highlight a late quaternary change in fluvial regime. *Australian Geographer, 40*, 29–49. https://doi.org/10.1080/00049180802656952

Wu, H., Yang, W., Yao, R., Zhao, Y., Zhao, Y., Zhang, Y., et al. (2020). Evaluating surface water quality using water quality index in Beiyun River, China. *Environmental Science and Pollution Research, 27*, 35449–35458. https://doi.org/10.1007/s11356-020-09682-4

Xu, H. (2006). Modification of normalised difference water index (NDWI) to enhance open water features in remotely sensed imagery. *International Journal of Remote Sensing, 27*, 3025–3033. https://doi.org/10.1080/01431160600589179

Yan, Q., & Ma, C. (2016). Application of integrated ARIMA and RBF network for groundwater level forecasting. *Environmental Earth Sciences, 75*. https://doi.org/10.1007/s12665-015-5198-5

Yang, X., Qin, Q., Yésou, H., Ledauphin, T., Koehl, M., Grussenmeyer, P., & Zhu, Z. (2020). Monthly estimation of the surface water extent in France at a 10-m resolution using Sentinel-2 data. *Remote Sensing of Environment, 244*, 111803. https://doi.org/10.1016/j.rse.2020.111803

Zemunac, R., Savic, R., Blagojevic, B., Benka, P., Bezdan, A., & Salvai, A. (2021). Assessment of surface and groundwater quality for irrigation purposes in the Danube-Tisa-Danube hydrosystem area (Serbia). *Environmental Monitoring and Assessment, 193*. https://doi.org/10.1007/s10661-021-09294-6

Zhang, G., Zheng, W., Yin, W., & Lei, W. (2020). Improving the resolution and accuracy of groundwater level anomalies using the machine learning-based fusion model in the North China Plain. *Sensors, 21*, 46. https://doi.org/10.3390/s21010046

Zhang, Q., Qian, H., Xu, P., Hou, K., & Yang, F. (2021). Groundwater quality assessment using a new integrated-weight water quality index (IWQI) and driver analysis in the Jiaokou Irrigation District, China. *Ecotoxicology and Environmental Safety, 212*, 111992. https://doi.org/10.1016/j.ecoenv.2021.111992

Zhou, Y., Li, P., Chen, M., Dong, Z., & Lu, C. (2020). Groundwater quality for potable and irrigation uses and associated health risk in southern part of Gu'an county, North China plain. *Environmental Geochemistry and Health, 43*, 813–835. https://doi.org/10.1007/s10653-020-00553-y

Chapter 2
Damodar Fan Delta in Context

2.1 Concept of Fan Delta

The term 'fan delta' (also termed fan-delta) is the combination of an alluvial fan and delta. Fan delta can be defined as an alluvial fan that progrades from an adjacent highland in standing waterbody either lacustrine or marine environment (Holmes, 1965; McGowen, 1971; McPherson et al., 1987). Wescott and Ethridge (1980) denoted that the highland adjacent to the coastline, the high gradient, and the bed-load stream that braided to the cost are the essential elements for the development of the fan delta. McPherson et al. (1987) restricted the term fan delta to the mass flow dominated dry type alluvial fan related to the tectonic escarpment and precluded the wet-type alluvial fan dominated by streams (Postma 1990). Nemec and Steel (1988) suggested that the fan delta should apply to all types of alluvial fans whenever actively in contact with sea or lake without considering whether it is gravelly or sandy, small or very large, dominated by streams or mass flows, and whether it is associated with escarpment or a virtually passive landscape. Therefore, the term fan delta can be defined as a prism of sediment delivered by an alluvial fan and deposited, mainly and entirely in subaqueous conditions, at the interface between an active fan and a standing body of water (Nemec & Steel, 1988). According to Nemec and Steel (1988), it is not necessary that always the alluvial fan from the highland comes into the contact with the standing waterbody by progradation. However, the sea or lake may encroach itself over an active fan and cause an interplay of the two environments, under the circumstance of an overall marine and lacustrine transgressions. The fan delta represents the interaction between subaerial and subaqueous environments represented by the sediment-laden alluvial fan system and standing waterbody, respectively. Hence, the fan delta includes the three essential attributes such as highlands, which are considered the source region of the sediments, sediment-laden alluvium fan system dominated by streams or mass flows as the feeder system, and standing waterbody into which the sediments are deposited

S. Mahammad, A. Islam, *Palaeochannels and Water Resource Management*,
https://doi.org/10.1007/978-3-031-92021-9_2

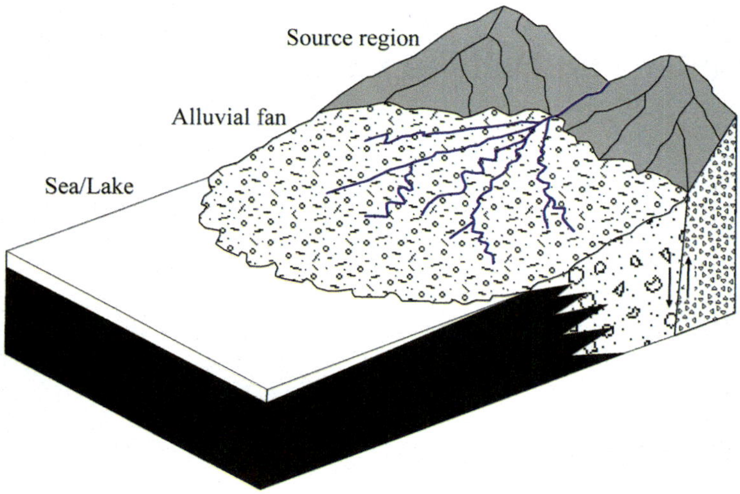

Fig. 2.1 Schematic diagram of fan delta (Based on McPherson et al., 1987)

(Fig. 2.1). The record of climatic change and sea level fluctuation is preserved in the fan delta in the form of distinct stratigraphic sequences (Postma & Middleton, 2003). In this study, the evolution of Damodar Fan Delta (DFD) has been analysed in terms of its subaerial and subaqueous configurations.

2.2 Evolution of Damodar Fan Delta and Palaeochannels

The evolution of the DFD started with the breakup of Gondwanaland and the sedimentation of the Bengal basin during the cretaceous periods (Alam, 1989, 1972). The study area experienced a series of marine transgression and regression since the Cretaceous and this may be attributed to regional tectonic uplift and associated subsidence (Singh 1998; Alam et al., 2003). In the initial phase, after crossing the Chotanagpur foothill escarpment, the Damodar River formed the Kuldiha formation adjacent to the hill area (Mahata & Maiti, 2019). This formation occurred under the continental framework of freshwater and fluviatile environments (Das, 1972). After the phase of Kuldiha formation, the eastern part of the area was inundated by the marine transgression of the Oligocene-Miocene period (Das, 1972). The western limits of the marine incursion are bounded by the NNE–SSW fault line (Sengupta, 1972; Das, 1972). During this period, the fan-forming process passes through the conditions of subaqueous and subaerial conditions (Mahata & Maiti, 2019). In the Miocene epoch, the sea level probably regraded and the fluviatile conditions were re-established in the area (Das, 1972). During this phase, the sedimentation process took place in the present-day adjoining area of Ajay and Damodar Rivers and formed the Alinagar formation, which comprises sand and pebbles alternating with grey clay and occasionally carbonaceous deposits (Das, 1972). The area experienced the last marine transgression in the early Pleistocene and after that present-day landform was established (Das, 1972).

The eastern part of the DFD is known as Memari Fan Delta which trends towards the east from its apex located at Barsul, near Barddhaman (Mallick & Niyogi, 1972). During the Eocene epoch, the highest marine transgression occurred which inundated the whole area of the Bengal shelf (Sengupta & Rao, 1966; Alam, 1989). In the Oligocene epoch, the sea level was situated below the Barddhaman (Sengupta, 1972). It is evident that during Oligocene, the deposition under freshwater to estuarine conditions continued near the Barddhaman, while shallow marine deposition took place at the Memari-Pandua area, far east of the Barddhaman (Sengupta & Rao, 1966). The marine transgression occurred from the late Miocene to Pliocene, which inundated the extensive area again (Sengupta, 1972). In the early Pleistocene, shallow marine conditions occurred in the deep basin area (Sengupta, 1972). Probably in the late Pleistocene, the sea finally receded completely from the Bengal basin and the erosion and peneplanation took place under the subaerial condition (Sengupta, 1972).

The formation of the delta, filled valley cuts, and river terraces concerning sea level oscillation during the quaternary period has been described by Mallick and Niyogi (1972) (Fig. 2.2). During the stage of sea level fall, a large section of the older plain was exposed into which the entrenched river flowed to meet the coastline at a right angle (Mallick & Niyogi, 1972). In this stage, the depositional part of the coastline moves towards the sea, and a river system carries coarser sediments due to high gradient. In the stage of sea level rise, the depositional zone of the delta is moving landward. In this stage, marine transgression occurs gradually and the exposed part of the delta comes in the soil oxidation conditions. During this stage, the river deposited coarser sediments from the substratum part of the delta and distributed the sediments by waves and currents of the rising sea. In the stationary stage of sea level, the river deposits sediments by vertical and lateral accretion. The deposition forms the top stratum with fine sediments. Mallick and Niyogi (1972) also described a landform developed by the rising and falling of sea level and they coined the term fill valley cut. During the low stand of the sea level, the linear trenches, running towards the sea are flowed by ancient rivers. These trenches are filled with sediments during the rising sea level. These linear valley-type landforms consist of sediments termed fill valley cuts (Fig. 3.3). The fill valley cuts are narrow towards the older plain and towards the sea marge gradually with the plain built by coalescing delta. The deposition of the filled valley cut is characterized by a gravel bed in the base grading up fine sand into silt and clay at the top. The filled valley cut represents the ancient buried channel (Mallick & Niyogi, 1972).

The coast of West Bengal experiences at least five glacial phases with alternative interglacial phases. Sea level oscillation has a significant effect on the development of the delta and associated landforms of the West Bengal coast. In the low stand of sea level, the ancient river of the coastal part flowed to the sea. In this stage, the river transported sediments to the coast. In the high stand of the sea level, the rivers deposited sediments and formed the ancient channel. After the regression of the sea,

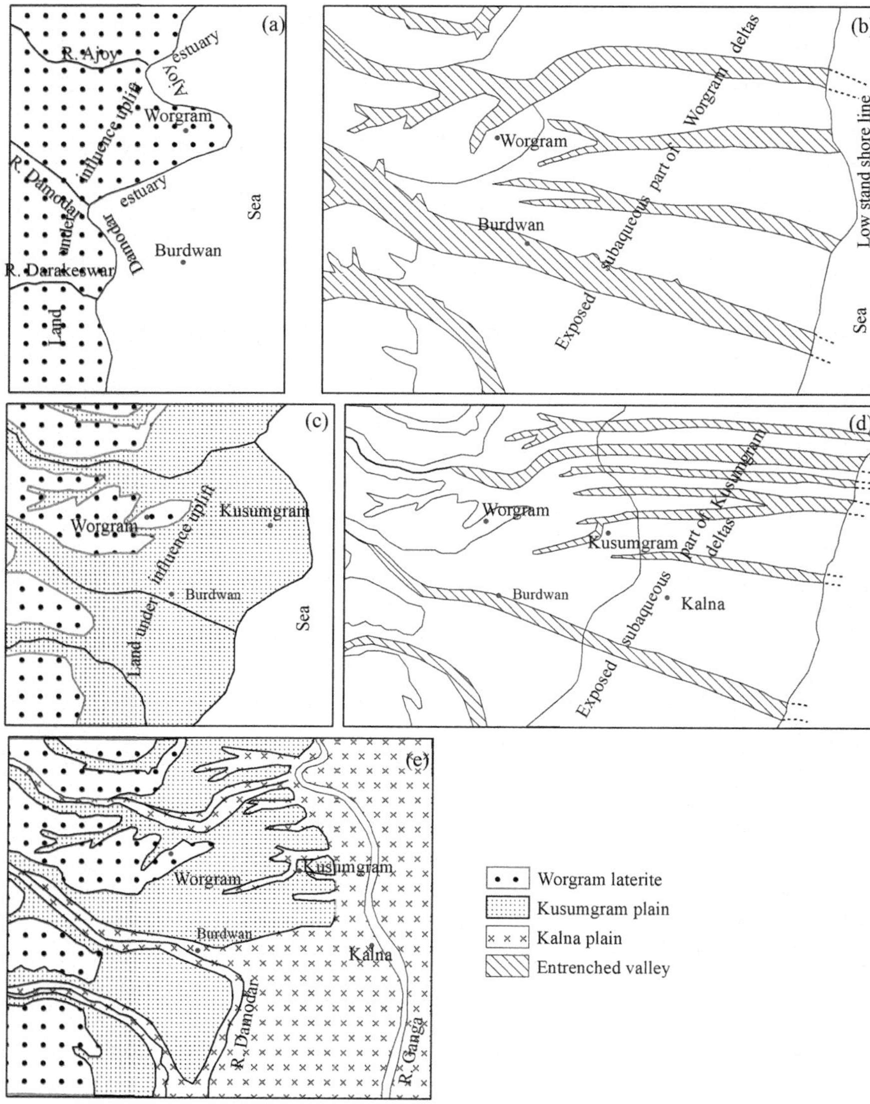

Fig. 2.2 Evolution of delta, filled valley cuts, and river terrace; (**a**) & (**c**) represent configurations during high stands, (**b**) & (**d**) represent configurations during low stands, (**e**) represents final configurations in the post Kalna time (Modified from Mallick & Niyogi, 1972)

the buried channels were concealed below the newer sediments. The authors examined the depositional sequence of some places in West Bengal with the help of borehole data. They found that at Barddhaman the bed of Damodar River is composed of a gravel layer of 240–250 m in depth and Pleistocene clay up to 240 m below the ground. In the well at Kalna, the thickness of the sand is more than 210 m. Therefore, it is evident that such a great thickness of sand gravel is not possible in a single-generation change, it has resulted from the multiple phases of filling, one coming above another (Mallick & Niyogi, 1972).

2.3 Physiography and Relief

The physiographical configuration of the South Bengal Basin can be grouped into six divisions. They are (a) degraded plateau located in the eastern fringe of Chaotanagpur plateau, (b) plateau fringe paleodeltas, (c) non-tidal upper Ganga delta, (d) tidal and reclaimed lower Ganga delta, (e) tidally inundated lower Ganga delta mainly occupied by Sundarbans mangroves, and (f) coastal plain of Medinipur (Bandyopadhyay et al., 2014).

The elevation of this region gradually drops to the east, which is controlled by three main faults such as the Chotanagpur Foothill Fault (CFF), the Medinipur Farakka Fault (MFF), and the Damodar Fault (DF) (Ghosh & Guchhait, 2019). The study area is located in the plateau fringe palaeodelta region and merges with the Ganga Delta to the east. The elevation of the study area ranges from 7 m near Amta (southern point) to 37 m near Barddhaman town (western point). Based on elevation, the study area is classified into six elevation zones such as 7–9, 9–11, 11–13, 13–15, 15–25, and 25–50 m (Fig. 2.3). The southern boundary of the study area is following the 7 m contour line, whereas the eastern boundary of the study area is tracing ~9 m contour line. The slope of the study area is almost gentle. The general slope of the study area trends towards the east and the southeast direction.

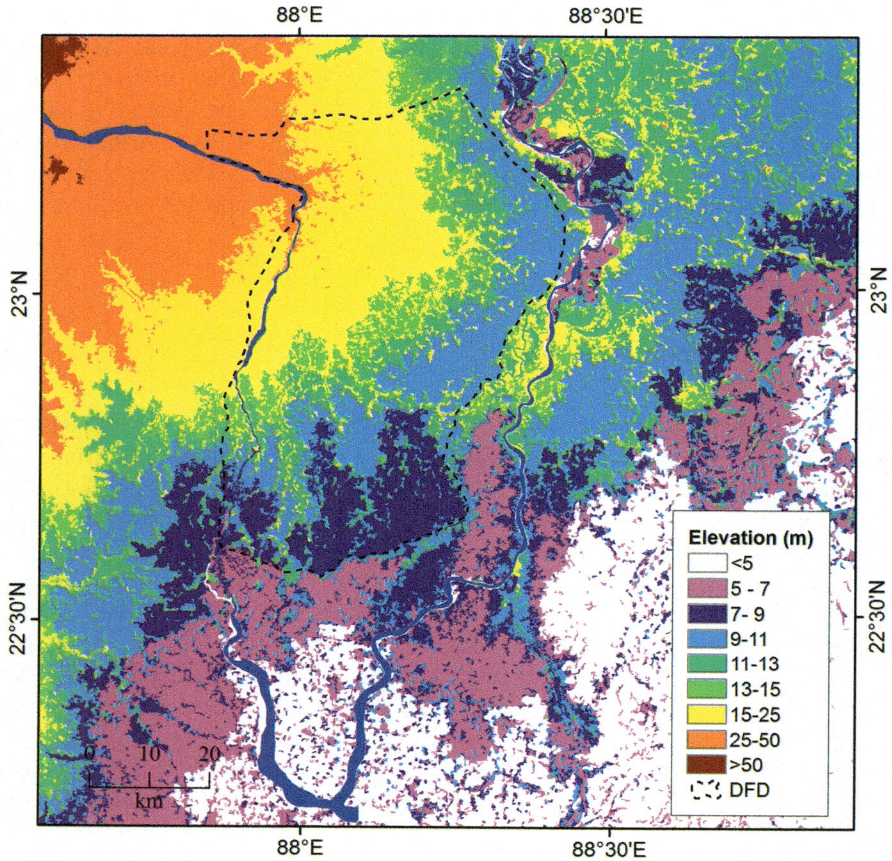

Fig. 2.3 Relief map of the study area and its surroundings (Source: SRTM DEM, 30 m)

2.4 Climate

It is observed that the study area has a tropical-humid to sub-humid climate characterized by a hot dry summer with low humidity (March–May), a hot wet monsoon with high humidity (June–November), and a short winter with low humidity (December–February) (Figs. 2.4, 2.5 and 2.6). The average temperature in the summer ranges from 30 to 40 °C, whereas in winter the average temperature ranges from 10 to 18 °C. May is the warmest month, whereas January is the coolest month in the year. The rainfall generally occurs in the monsoon period on an average of 1550 mm annually. The rainfall in this area is evenly distributed at certain intervals due to the abrupt nature of the monsoon. In this area, more than ~70% of rainfall occurs in the monsoon period and ~30% of the rainfall occurs rest of the period. Similar climatic conditions of the study area have been reported by Bhattacharyya (2011).

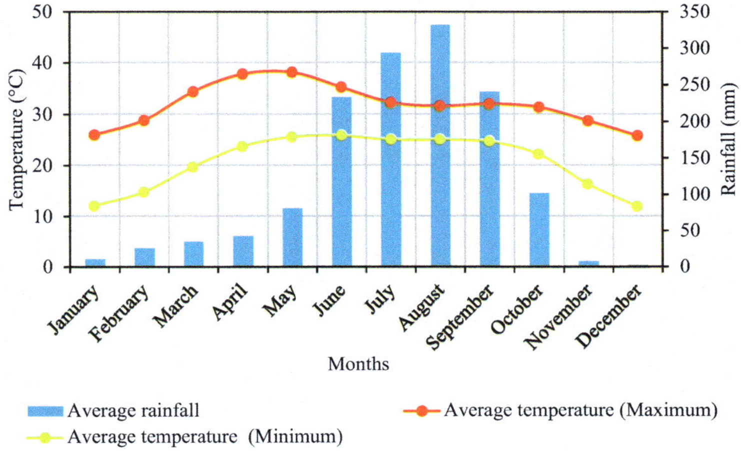

Fig. 2.4 Temperature and rainfall distribution of Barddhaman district (Computed from India WRIS, 1901–2002)

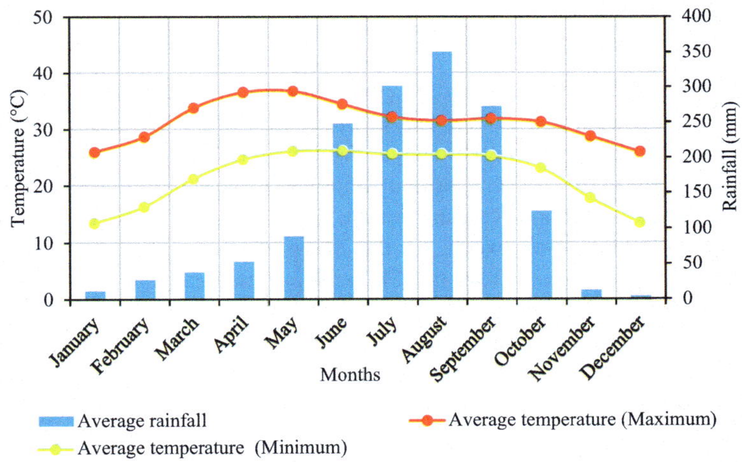

Fig. 2.5 Temperature and rainfall distribution of Hooghly district (Computed from India WRIS, 1901–2002)

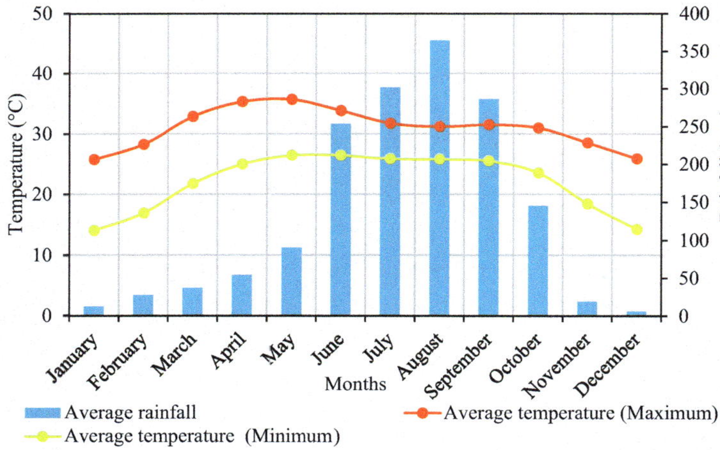

Fig. 2.6 Temperature and rainfall distribution of Howrah district (Computed from India WRIS, 1901–2002)

2.5 River System

West Bengal is a riverine state and its environment, society, culture, and economy are greatly influenced by the rivers (Bandyopadhyay, 2007). The major rivers of the state originate either from the Himalayan Mountain or from the Chotanagpur plateau. In the central and western parts of South Bengal, Bhagirathi-Hooghly, Ajay, Damodar, Darakeswar, and Rupnarayan are the important rivers. The rivers Ajay, Damodar, and Darakeswar, except for the Bhagirathi-Hooghly, are rainfed in this area because they originate from the Chotanagpur plateau. Therefore, the flow of the rivers oscillates with the intensity of rainfall, even causing devastating floods in the monsoon season. The study area is bounded by the Hooghly River to the east and the Damodar River to the west. The study region is traversed by the Damodar River and its numerous disconnected courses. The river originates from the Chotanagpur plateau, travels through the states of Jharkhand and West Bengal for about 540 km, and then merges with the Rupnarayan River to discharge into the Hooghly River (Rudra, 2010).

The catchment area of the study river is 220,069 km^2, of which 5250 km^2 is in West Bengal (Rudra, 2010). Since ancient times, the Damodar River has been accountable for colossal floods. In its lower portion, the river has altered its course multiple times throughout history. According to earlier studies, old maps, and historical data, the Damodar River used to flow through Khari, Banka, Behula-Gangur, Kana-Kunti, and Kana Damodar at different times. In places where these channels appear as small waterbodies and marshes, they are dried up, filled with sand, and transformed into DVC canals (Fig. 2.7). Buried and abandoned channels can be useful in the artificial recharging of excessive floodwater in this study area. Polley et al. (2022) demonstrated that the study area consists of several abandoned channels, palaeolevees, and scars which are evidence of channel avulsion.

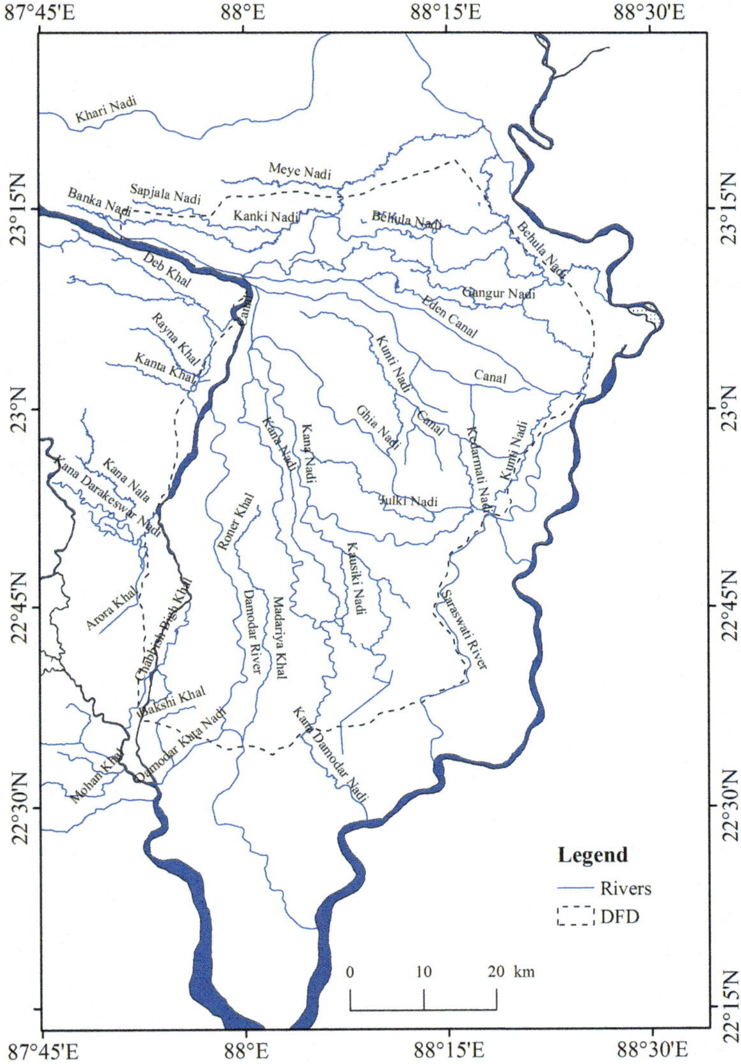

Fig. 2.7 Drainage map of the study area (Prepared by the authors, 2020 based on toposheets)

2.6 Lithology and Soil

The study area is covered by the thick alluvium of the Holocene to the upper Pleistocene epoch (Fig. 2.8). The alluvium consists of sand, silt, clay, clay alternating with sand and silt, and often clay with caliche concentrations of the Hugli (also termed Hooghly), Panskura (Chinsurah/Katwa), and Sijua formation, respectively (GSI, 2001, 2002, 2006). Hugli formation is the present-day deposits of rivers characterized by unoxidized and less compact grey to pale yellowish sediments. It is found in the floodplains of the Hooghly, Damodar, and Tarakeswar rivers of the study area. Panskura formation is constituted by clay with alteration of silt and fine sand and formed the para-deltaic fan surface of the study area. It covers the largest part of the study area. Sijua formation is the oldest alluvium sequence characterized by oxidized red to brownish clay with caliche concentrations found in the western and eastern flanks of the study area. Moreover, the location-specific detailed subsurface profiles are discussed in Chap. 3 (Sect. 3.4). The soil types of the study area are heterogeneous in terms of depth, texture, the requirement of irrigation, and cropping pattern. The major five types of soil classes and their association such as Haplaquepts, Ustochrepts, Ustifluvents, and Ochraquults are confined in the study area (NBSS & LUP, 1992) (Fig. 2.9). The Haplaquepts type is characterized by very deep, poorly drained, fine soil with a clay surface. This type of soil is found in the low-lying alluvial plain of the study area.

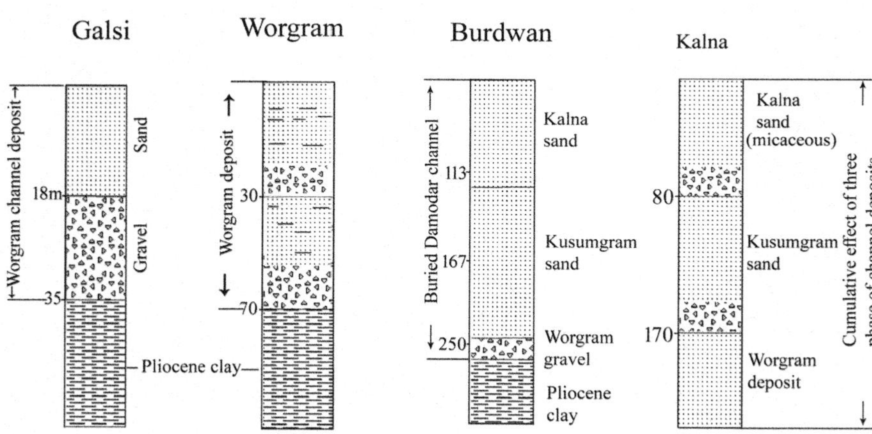

Fig. 2.8 Lithological composition of the Damodar Fan Delta (Source: Mallick & Niyogi, 1972)

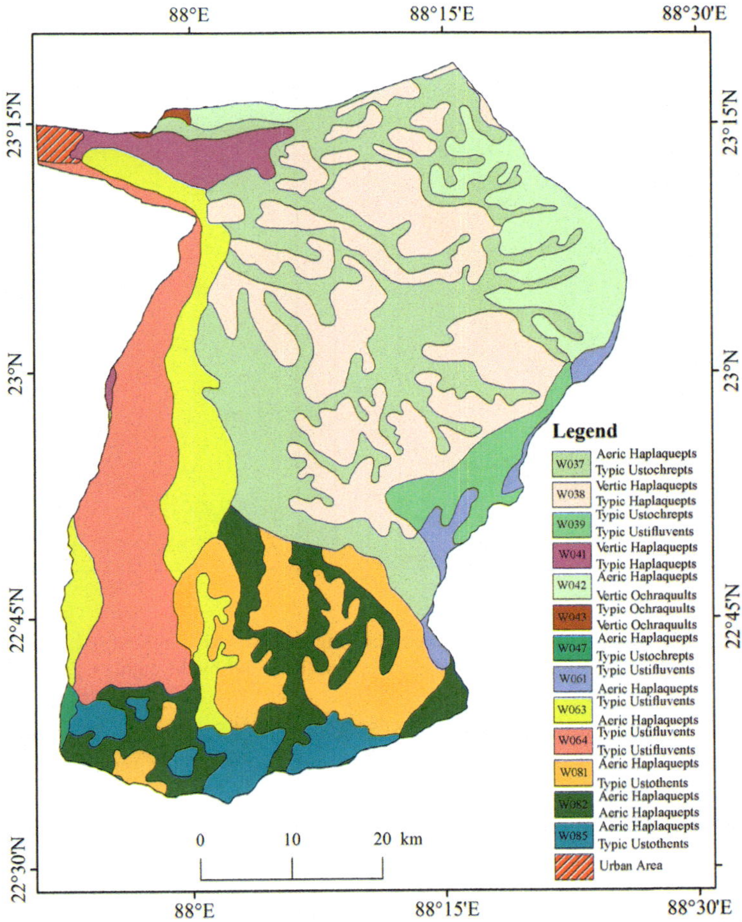

Fig. 2.9 Soil map of the study area (Prepared from NBSS & LUP, 1992)

2.7 Land Use and Land Cover

The supervised image classification is accomplished to show major land use and land cover (LULC) classes in the pre-monsoon and post-monsoon periods of the DFD for the recent time (2020). The study area consists of six LULC classes such as (1) waterbody, (2) vegetation cover, (3) agricultural land, (4) agricultural fallow land, (5) sandy area, and (6) built-up area of pre-monsoon (Fig. 2.10a) and the post-monsoon period (Fig. 2.10b). The waterbody consists of an area of 96.76 km^2 in the post-monsoon season and 85.38 km^2 in the pre-monsoon season due to the seasonal distribution of precipitation. It includes all the open waterbody such as rivers, ponds, bills, and canals. The vegetation type of the study area is deciduous forest, shrubs,

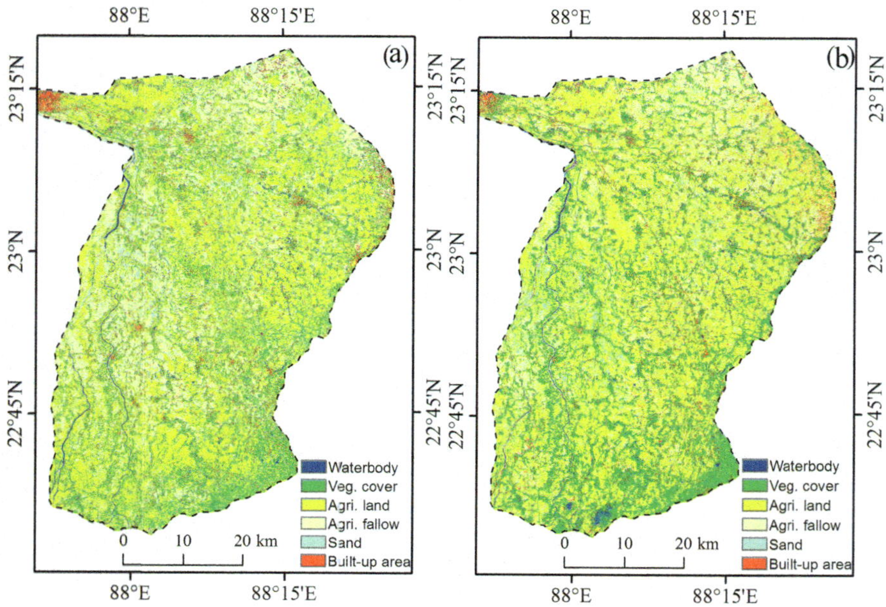

Fig. 2.10 LULC map of the DFD for 2020 showing the dynamics in land use and land cover classes for (**a**) pre-monsoon, (**b**) post-monsoon season (Prepared by the authors, 2022)

grassland, and plantation, which covers 591.02 and 685.67 km^2 of the study in the pre-monsoon and post-monsoon periods, respectively. Vegetation cover has appeared in a linear pattern followed by the relict, spill, and abandoned channel, marshes, and canals of the study area. The area is dominantly covered by agricultural land, which accounts for 1523.10 km^2 in the post-monsoon season and 1420.16 km^2 in the pre-monsoon period. Rice is the dominant crop in the study area and the Purba Barddhaman district located in the study area is known as the 'Bowl of Rice' for huge rice cultivation. The rice is grown seasonally in the form of *Aus* (autumn), *Aman* (winter), and *Boro* (spring) based on the rainfall and type of agricultural land. Apart from that, jute is an important crop preferably yielding in low-lying areas. Pulses like gram, *khesari*, peas, and *musuri* are also cultivated. In addition, potatoes, onions, vegetables, and oilseeds are cultivated in this area. In the pre-monsoon period, the cultivated area is reduced due to a lack of rainfall. In the lean period cultivation, especially for *boro* rice is done by irrigation from a deep tube well, shallow tube well, and dug wells. Therefore, agricultural fallow is higher than in the post-monsoon period. The agricultural land consists of both cropped and uncropped areas.

The agricultural lands are found throughout the study area surrounding the settlement and are more intensive along the riverbank of the study area. The vegetation cover is predominantly followed by agricultural land. Besides, the sand area of LULC consists of a channel bar of the river and bared land with sandy exposures of the study area. It fluctuated from 41.99 to 19.51 km^2 from the post-monsoon to the pre-monsoon period. In addition, the built-up area covers 110.26 and 92.43 km^2 in the post-monsoon and pre-monsoon seasons, respectively. It includes urban and rural settlements, railways and roads, and structure to livelihood. The transport network of the study area includes Howrah to Barddhaman main and chord railway line, Howrah to Katwa railway line, Howrah to Tarakeswar railway line, and Howrah to Kharagpur railway line, Grand Trunk (GT) road, Delhi road, and state highway which connect the study area with the state.

2.8 Demography and Society

The demographic and sociocultural investigation of the study area is required to comprehend their role in effectively managing the available resources for development. The parts of Purba Barddhaman, Hooghly, and Howrah districts collectively cover the entire study area, which consists of 29 CD blocks, with 9, 15, and 5 CD blocks coming under each district, respectively. The total population under the CD blocks of these three districts, according to the census 2001 and 2011 are 5,913,417 and 6,518,179, respectively, and therefore increased by 10.22%. The number of households reached 1,494,118 and the female–male ratio is recorded as 96.62 as per the census 2011. The share of scheduled caste and scheduled tribes population to total population is 28% and 6%, respectively.

The enhancement of education acts as the main driver for the continued development of human resources and is also a critical element of the nation's socioeconomic well-being. A rise in literacy is necessary to improve quality of life. The average literacy percentage in this area is 78.76% with 42.97% male literacy and 35.79% female literacy. Regarding the working population, the total working population has been registered as 2,639,525, where the share of the male is ~77%. Besides, in terms of the human development index in 2011, the Howrah district ranked 2 followed by the Barddhaman (rank 5) and Hooghly (rank 6) districts in West Bengal. Among the CD blocks of the Barddhaman district located in the study area, Barddhaman I and Barddhaman II CD blocks comprise the higher HDI (0.54 and 0.53, respectively) (GoWB, 2011). In the Hooghly district, Singur ranked first (0.63). Besides, Chanditala I and II and Chisurah-Mogra hold the second and third positions, respectively. Moreover, Balagarh, Polba-Dadpur, Khanakul I and II, and Pursurah CD blocks account for the lower HDI indicating the poor condition of human development in the district (GoWB, 2010).

2.9 Physico-Social Adversities

There are several physical adversities in the study area. Physical and social hazards contribute to the instability of the socioeconomic and physical aspects of the study area. The major natural hazards that occur in the study area are flood, drought, cyclone, depletion of groundwater level, and deterioration of groundwater quality, especially arsenic contamination. The flood affects the CD block located along the Damodar River in the lower part of the study area. The abrupt nature of the monsoon spell, funnel shape river basin, and sudden release of water from the DVC reservoir is responsible for the flood condition of the study area.

Apart from floods, lack of monsoon rainfall causes meteorological drought, which induces hydrological, agricultural as well as socioeconomic drought in the study area (Datta et al., 2022). Besides, the study also found tropical cyclone, which causes water-logging and damage to crops and properties. Moreover, the depletion of groundwater level is a serious constraint in the study area, which results from the overexploitation of groundwater for irrigation and industrial uses (Majumder & Sivaramakrishnan, 2014). The study area is agriculture dominated and popular for growing water-intensive crops such as *boro* rice (Ray & Ghosh, 2007). Therefore, it accelerates the water scarcity condition further. The depletion of groundwater level is associated with arsenic contamination, especially in the recent alluvial belt (Pal & Mukherjee, 2010). Additionally, the use of fertilizers, pesticides, and herbicides in agricultural land induces problems regarding water contamination.

With this backdrop, the next chapter would delve into the delineation and characterization of the palaeochannels for potential water resource management for this agro-based region.

References

Alam, M. (1972). Tectonic classification of Bengal basin. *Geological Society of America Bulletin, 83*(2), 519–522.

Alam, M. (1989). Geology and depositional history of Cenozoic sediments of the Bengal Basin of Bangladesh. *Palaeogeography, Palaeoclimatology, Palaeoecology, 69*, 125–139.

Alam, M., Alam, M. M., Curray, J. R., Chowdhury, M. L. R., & Gani, M. R. (2003). An overview of the sedimentary geology of the Bengal Basin in relation to the regional tectonic framework and basin-fill history. *Sedimentary Geology, 155*(3–4), 179–208.

Bandyopadhyay, S. (2007). Evolution of the Ganga Brahmaputra delta: A review. *Geographical Review of India, 69*(3), 235–268.

Bandyopadhyay, S., Kar, N. S., Das, S., & Sen, J. (2014). River systems and water resources of West Bengal: A review. *Geological Society of India Special Publication, 3*(2014), 63–84.

Bhattacharyya, K. (2011). The Lower Damodar River, India: understanding the human role in changing fluvial environment (p. 295). New York: Springer.

Das, S. (1972). Sedimentation in the Ajay—Damodar Basin. In K. G. Bagchi (Ed.), *The Bhagirathi—Hooghly Basin* (pp. 9–17). University of Calcutta.

Datta, D., Banerjee, S., Kar, G., Ghosh, S., & Saha, S. (2022). Spatio-temporal dynamics of temperature and rainfall across jute growing districts of India. *Mausam, 73*(2), 373–388.

Ghosh, S., & Guchhait, S. K. (2019). Modes of formation, Palaeogene to early Quaternary palaeogenesis and geochronology of laterites in Rajmahal basalt traps and Rarh Bengal of Lower Ganga Basin. *Quaternary Geomorphology in India: Case Studies from the Lower Ganga Basin*, 25–60.

Government of West Bengal (GoWB). (2010). *District human development report—Hooghly. West Bengal.* Development & Planning Department, Government of West Bengal.

Government of West Bengal (GoWB). (2011). *District human development report—Bardhaman. West Bengal*. Development & Planning Department, Government of West Bengal.

GSI. (2001). *District resource map, Barddhaman, West Bengal*. Geological Survey of India.

GSI. (2002). *District resource map, Howrah, West Bengal*. Geological Survey of India.

GSI. (2006). *District resource map, Hooghly, West Bengal*. Geological Survey of India.

Holmes, G. W. (1965). *Geologic reconnaissance along the Alaska Highway, Delta River to Tok Junction, Alaska*. US Government Printing Office.

Mahata, H. K., & Maiti, R. (2019). Evolution of Damodar Fan Delta in the Western Bengal Basin, West Bengal. *Journal of the Geological Society of India, 93*, 645–656.

Majumder, A., & Sivaramakrishnan, L. (2014). Ground water budgeting in alluvial Damodar fan delta: A study in semi-critical Pandua block of West Bengal, India. *International Journal of Geology, Earth & Environmental Sciences, 4*(3), 23–37.

Mallick, S., & Niyogi, D. (1972). Morphology of the Midnapore district, West Bengal. In *Proc. Seminar Geomorphology, Geohydrology and Geotectonic of the Lower Ganga Basin, Kharagpur: A81 A* (Vol. 86).

McGowen, J. H. (1971). *Gum Hollow fan delta, Nueces Bay, Texas*. Virtual Landscapes of Texas.

McPherson, J. G., Shanmugam, G., & Moiola, R. J. (1987). Fan-deltas and braid deltas: Varieties of coarse-grained deltas. *Geological Society of America Bulletin, 99*(3), 331–340.

NBSS, & LUP (1992). Soil of West Bengal for optimizing land use. *Indian Council of Agricultural Research*, Nagpur.

Nemec, W., & Steel, R. J. (1988). What is a fan delta and how do we recognize it. *Fan Deltas: Sedimentology and Tectonic Settings, 3*(13), 231–248.

Pal, T., & Mukherjee, P. K. (2010). Search for groundwater arsenic in Pleistocene sequence of the Damodar River flood plain, West Bengal. *Indian Journal Of Geosciences, 64*(1–4), 109–112.

Polley, P.K., Saha, U.D., Sen, A. (2022). Anthropogeomorphic Understanding of the Damodar Fan Delta, India. In: Islam, A., Das, P., Ghosh, S., Mukhopadhyay, A., Das Gupta, A., Kumar Singh, A. (eds) Fluvial Systems in the Anthropocene. Springer, Cham. https://doi.org/10.1007/978-3-031-11181-5_16.

Postma, G. (1990). Depositional Architecture and Facies of River and Fan Deltas: A Synthesis. *Coarse-grained deltas*,13–27. https://doi.org/10.1002/9781444303858.ch2.

Postma, G., & Middleton, G. V. (2003). Fan delta. *Encyclopedia of sediments and sedimentary rocks*. Kluwer Academic Publishers, Boston, 272–274.

Ray, S., & Ghosh, D. (2007). Modern agriculture and the ecologically handicapped: Fading glory of boro paddy cultivation in West Bengal. *Economic and Political Weekly*, 2534–2542.

Rudra, K. (2010). Dynamics of the Ganga in West Bengal, India (1764–2007): Implications for science–policy interaction. *Quaternary International, 227*(2), 161–169.

Sengupta, S. (1972). Geological framework of the Bhagirathi-Hooghly basin. *The Bhagirathi-Hooghly Basin*, 3–8.

Sengupta, S., & Rao, J. S. (1966). Statistical analysis of cross-bedding azimuths from the Kamthi formation around Bheemaram, Pranhita: Godavari Valley. *Sankhyā: The Indian Journal of Statistics, Series B*, 165–174.

Singh, L. P., Parkash, B., & Singhvi, A. (1998). Evolution of the Lower Gangetic Plain landforms and soils in West Bengal, India. *Catena, 33*(2), 75–104. https://doi.org/10.1016/S0341-8162(98)00066-6.

Wescott, W. A., & Ethridge, F. G. (1980). Fan-delta sedimentology and tectonic setting—Yallahs fan delta, Southeast Jamaica. *AAPG Bulletin, 64*(3), 374–399.

Chapter 3
Identification and Characterization of Palaeochannels

3.1 Etymology and Concept of Palaeochannels

Palaeochannels (PLC) (also spelled paleochannel) are an important geomorphic feature in the riverine floodplain from environmental as well as economic perspectives. PLC is also known as the old river, lost river, buried channel, or buried valley. The word PLC is derived from two words 'palaeo' or 'old' and channel, i.e. PLC is an old channel. A pile or sheet of younger alluvium packed with the remains of former streams and channel courses, or PLC, is a representation of the historical fluvial system (Rajawat et al., 2003; Rathore et al., 2010). PLC occurs when a portion of the active river system ceases from the water flow and is filled completely or partly with recent sediment and transformed into a buried, lost, or shifted river system due to tectonic, geomorphic, climatic as well as anthropogenic processes or activities (CGWB, 2016). Hence, according to Page et al. (1996), palaeochannels can be located at the modern land surface or in the subsurface (Harris, 1980). The fills of the PLC mainly contain unconsolidated and semi-consolidated sediments (CGWB, 2016). Hence, it is considered a subsurface conduit of the groundwater flow even if it causes a high risk of irrigation water losses by deep drainage (Wray, 2009). The geometry of the channel and its location on the floodplain in relation to overbank occurrences determine the infill rate of PLC deposits (Charlton, 2007). PLC can be observed as fluvial landforms in elongated, oblate, and sinuous shapes, which reveals the drainage system of the geological past (Mohammed-Aslam & Balasubramanian, 2010). The spatial extension of the PLC ranges from a small meander cut-off to an entire channel or channel belt expanding over tens to hundreds of kilometres (CGWB, 2016). It is obvious that the PLC does not always flow water for a long period except perhaps in the peak flood season (Wray, 2009). The deposits of PLC differ from the presently active river because it contains porous sedimentary fills. Water-bearing porous gravels, fine to medium sand (in river channel zones), silt (in delta regions), and coarse to medium sand (in alluvial fans) can

S. Mahammad, A. Islam, *Palaeochannels and Water Resource Management*,
https://doi.org/10.1007/978-3-031-92021-9_3

all be found in the palaeochannel. Due to the high infiltration capacity, PLC is considered a potential source of groundwater recharge and it serves as a flood channel and aqueduct.

3.2 Palaeochannel Identification Techniques

The identification of PLC can be done through historical records, locations of archaeological sites, surface and near-surface expression using remote sensing data, and sub-surface configurations using geological data analysis. In general, there are three broad methods for the identification of palaeo-channels: (1) historical maps and records, (2) geo-physical and geomorphological survey, and (3) geo-spatial techniques. The historical maps, archives, and other documentary records can identify palaeochannels although with some analogies and approximations. Geo-physical survey techniques are modern techniques usually done with ground penetrating radar (GPR) and resistivity meter. These are in-depth and accurate techniques, but they are often costly, hence limited to spatial and temporal extent. Under this category, geomorphological survey based on the surface cross-sections and sub-surface expression of the landforms/structure, palaeochannels can be identified. Third category is low cost identification of palaeochannels using remote sensing and GIS techniques. However, field validation of this technique is required to ensure the nature and properties of the palaeochannels. In this chapter, the identification of PLC has been done using historical maps and documents, literary records, lithological data analysis, and remote sensing techniques. Because of the fragmentary nature of data in space and time, combinational and integrated methods are essential. For example, for the historical past, remote sensing and geo-physical are not available. Thus, this chapter is concerned with the integrated approach, describing data acquisition from historical literature, lithology, satellite images and integrating all of them to get a result.

3.3 Identification of Palaeochannel Using Historical Documents and Literary Records

Historical documents provide significant information about sequential changes in the river courses. Based on the historical documents and maps, Ghosh and Mistri (2012), Bhattacharyya (2011), and Chakraborty et al. (2011) identified the ancient courses of the Damodar River. The Damodar River, popularly known as the 'Sorrow of Bengal' is an important western tributary of the Ganga River (Rudra, 2010). Starting on the Chotanagpur Plateau, it travels over 540 km across the states of West Bengal and Jharkhand until joining the Hooghly River as the Rupnarayan River's combined flow (Rudra, 2010). The river is renowned for its abrupt lower course change, which alters the topography of the districts of Barddhaman, Hooghly, and Howrah (Ghosh & Mistri, 2012). The past courses of the Damodar River are completely different from the present channels. The old channels of the Damodar River can be extracted with the help of precious old maps, charts, and literature records. In this regard, the maps of Jao de Barros (1550), Van Den Broucke, Thronton (1675), Cantelli da Vignolla (1683), Rennel (1776), and Walker (1867–1871) are considered cognizable shreds of evidence (Mukerjee, 1938). According to Jao de Barros (1550), the sixteenth-century main flow of the Damodar followed the path of the now-degraded Kana Damodar, which split below Selimabad and debouched into the Hooghly near Uluberia, as documented by Stevenson and others (Biswas, 2001a). Van Den Broucke's map (1660) demonstrates how, in the seventeenth century, the primary Damodar River flow was maintained through Moja Damodar and found to fall into the Rupnarayan River close to the current Bakshi Khal (Fig. 3.1a). Simultaneously, the larger Damodar branch used to flow through Barddhaman in an easterly direction, following the Gangur River's straight course until it met the Hooghly River close to Kalna. The Saraswati River flowed from Tribeni to the Hooghly River and was depicted on this map as an active river. Stevenson and others said that the Kalna branch had dried up (Biswas, 2001a). Simultaneously in 1660, a different branch of the Damodar reached the Hooghly River across Falta, via the Amta Canal. Mondal Ghat River was the name given to it (Bhattacharyya, 2011). It was discovered that the Kana Damodar, which at that time was the primary Damodar channel, entered the Hooghly River close to Uluberia. Within the Thornton chart (1675), a similar alignment of the river system of Van den Broucke is observed. Therefore, at the end of the seventh century or early eighteenth century, Kana and Kunti Nadi have a new channel available. This took off near Selimabad at the great bend of the Damodar and flew east along the current Kana Nadi route near Gopalnagar. After there, it turned northeast along the current Kunti Nadi and reached the Hooghly River 4.8 km upstream of Tribeni near Noaserai (Bhattacharyya, 2011). In the Rennell map (1776), the main flow is confined in the Amta channel and meets the Hooghly River through the Mangalghat River (Fig. 3.1b). It is also observed that the eastward channel of the Damodar River appears as a small channel through the Banka *Nadi*. In this map, the Kana-Kunti River taken off from Selimabad flows through the places of Parambua, Dipa, Dwarhatta, and Bandipur and is observed to fall into the Hooghly River at Noaserai.

According to Stevenson and others, a chart from 1690 depicts the Kana Damodar as a sizable torrent that enters the Hooghly River close to Uluberia and is known as the Jon Perdo River (Biswas, 2001a) (Fig. 3.2). The Kana and Kunti Nadi dried up about 1770, and the major flow went through the Mundeswari River (Stevenson and others), as well as the Amta Canal (Bhattacharyya, 2011). The Amta Canal and Mundeswari River serve as the Damodar River's primary flow now, while the river's historic routes can still be seen as PLC, spill channels, remnant channels, and abandoned channels. Notable examples are the Khari, Behula-Gangur, Ghia-Kantal, Kana-Kunti, and Kana Damodar.

Ancient literary records provide important evidence to reconstruct the period in which it was written. *Mangal Kavya* of the Bengali literature accounts for the

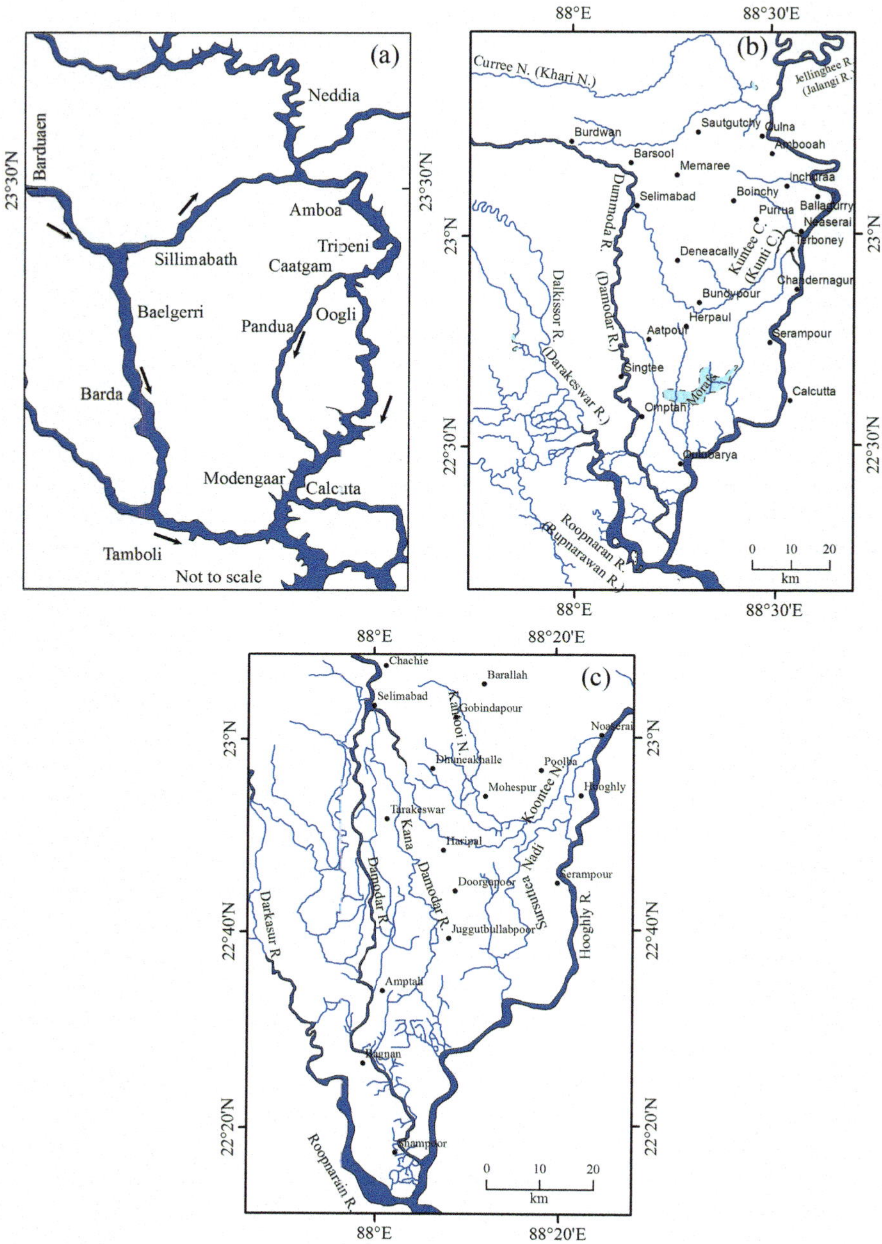

Fig. 3.1 Changing courses of Damodar River: (**a**) Van Den Broucke's map (1660), (**b**) Rennell (1776), and (**c**) Walker (1867–1871) (Fig. 3.1a, b based on Mukerjee (1938) and Fig. 3.1c based on Ghosh (2011))

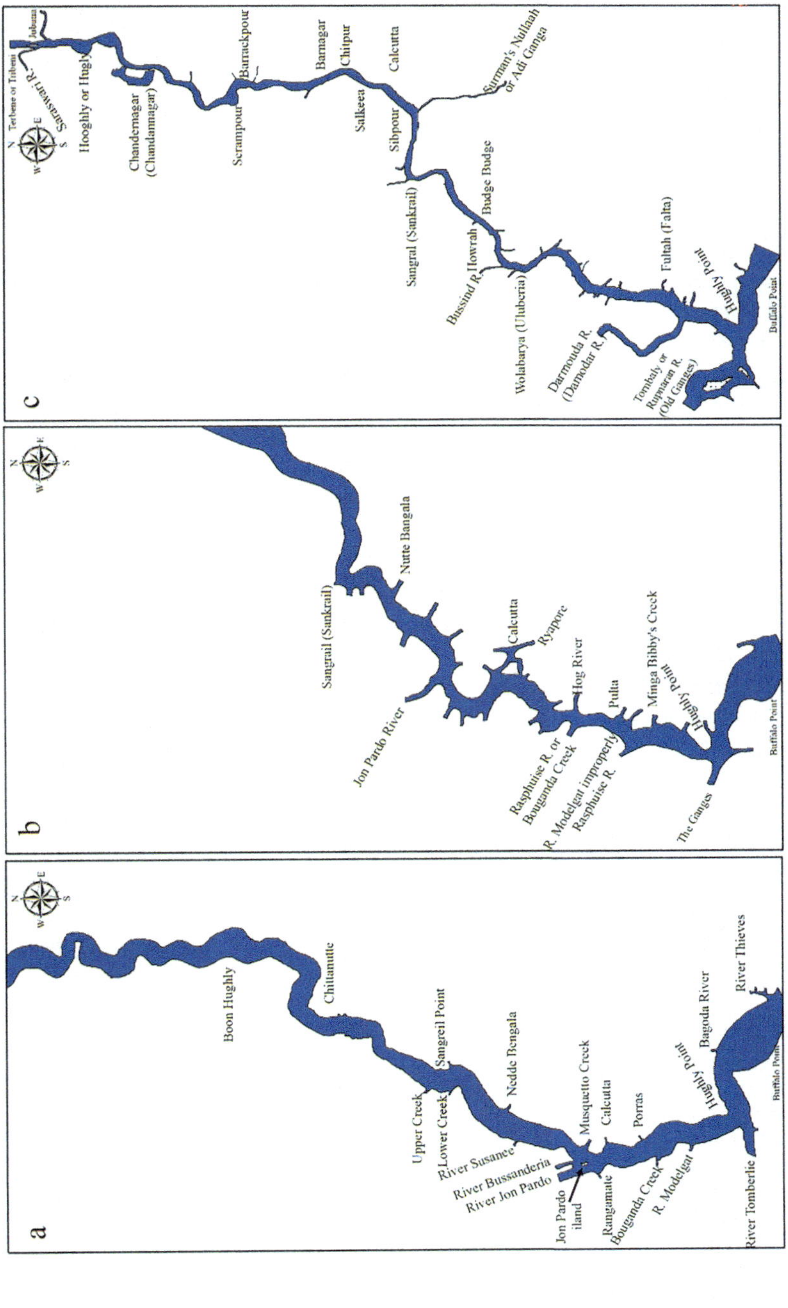

Fig. 3.2 Evolutions of Kana Damodar and Mangalghat River, (**a**) 1690, (**b**) 1720, (**c**) 1785 (Based on Biswas, 2001b)

detailed information about geographical descriptions, history, and social system of medieval Bengal, especially of Rarh Bengal. The *Mangal Kavya* written from the thirteenth century to the eighteenth century gives testimony to the former geographical description of different places. Based on the literary evidence of *Mangal Kavya*, many researchers delineated the old river courses. Bandyopadhyay (1996) detected the *Adi Ganga* PLC in the south 24 Parganas based on *Mangal Kavya* of the medieval period. Ghosh and Jana (2019) identified the ancient path of the Damodar Rivers based on the locations described in *Manasa Mangal Kavya. Manasa Mangal* written by Ketakadas Kshemananda is the oldest *Mangal Kavya* and deals with the establishment of the Godhead of *Manasa*, a Goddess of snakes, in the mortal. It is believed that *Manasa* chose *Chandsadagar*, the ardent devotee of Lord Shiva to acknowledge *Manasa's* divinity (Ghosh & Jana, 2019). *Chandsadagar*, a rich merchant of Champaknagar refused to confirm her godhead and engaged in conflicts with goddess *Manasa. Manasa* takes revenge upon Chandsadagar by destroying his fourteen ships with his acquired wealth and killing his six sons. *Manasa* also killed his seventh son *Lakhindar* on his wedding night. *Behula*, the newly wed wife of *Lakhindar*, decides to sail with his dead body to restore his life. After sailing for 6 months along the Gangur River, she reached the abode of gods and brought back Lakhindar alive. On the way to heaven and returning home, *Behula* described many places, which indicate the course of the Gangur River. In this chapter, some places mentioned in the *Manasa Mangal Kavya* have been found to define the ancient course of the Damodar River, such as Champai (Champaknagar), Gobindapur, Goda (Goda Ghat), Barddhaman, Gangpur (Gangapur), Amadpur, Narkeldanga, Hasanhati, Baidyapur, Gaharpur, and Tribeni (Fig. 3.3).

The *Dharma Mangal Kavya* narrates the story of the *Puranic* king *Harishchandra*, his queen *Madana* and his son *Luichandra* and another part describes the story of *Lousen*. The description of *Dharma Mangal Kavya* provides information regarding the geographical description of medieval Bengal. The important places mentioned in *Dharma Mangal* by Manik Ram Ganguli are *Simul Nagar*, Basdinga Garh, Bahirkhanda, and Bandipur. *Simul Nagar* was the ancient kingdom of King *Haripal*, which is presently located in the Haripal of Hooghly district. Manik Ram Ganguly, the author of the *Dharma Mangal*, described the location of *Simul Nagar* as,

Sakksaṯ Sōnār Laṅkā Simul Nagar.
Brāhmaṇī Bēṣṭita Tār Jemon Sāgar

[Meaning: *Simul Nagar* is a prosperous place as Lanka which is bounded by Brahmani River such as a sea.]

Another author of the *Dharmamanga Kavya* noted that *Simul Nagar* located along the Bimala River.

Dibas Rajani Chalē Nahi Raye Sthir.
Simul Samipe Gela Bimalar Tir

[Meaning: Days and nights go continuously that are not fixed. The bank of the Bimala River shallowed near *Simul*.]

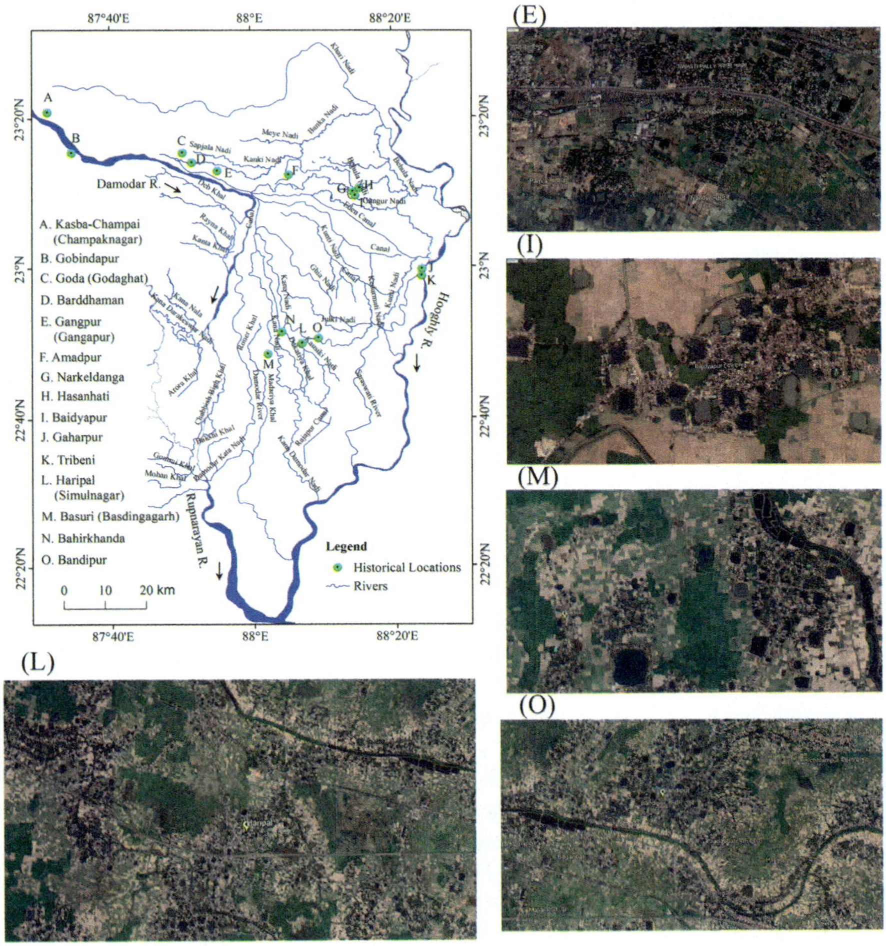

Fig. 3.3 Ancient course of Damodar River defined by the places mentioned in *Mangal Kavya* (Image source: Google Earth images, date 23 April 2022)

Therefore, it is observed that *Simul Nagar* was bounded by two rivers, Bamhani and Bimala Rivers. It also depicts that the rivers were large in dimension. At present, no large river has been found around the Haripal area. It is also mentioned that Haripal is presently flanked on the northern side by the Kana River flowing west to the east direction and the Koushiki River on the southern side flowing north–south direction (Fig. 3.3L). It is inferred that the Kana and Koushiki Rivers were broad rivers in the medieval period and now transformed into seasonal flows which are confined to a narrow channel.

In *Dharma Mangal*, Sahadev Chakraborty describes the location of the Bandipur beside the old Damodar River.

Bandipurer Bandiba Thakur Shymrai.
Damodar Jahar Dakṣhiṇe Baai.

[Meanings: Thakur Shymrai is adorated in Bandipur, which is surrounded by Damodar River flows in the southern part.]

It indicates that the Damodar River was confined to the southern part of Bandipur in the medieval period. It flowed towards the east passing through the Parambua, Dipa, Dwarhatta, and Bandipur. At present, Bandipur is located in the northern part of the Kana *Nadi* (Fig. 3.3O). Therefore, the Kana *Nadi* is known as the Damodar River and now it is the ancient path of the Damodar River.

Ghanaram Chakraborty noted that *Dharmapala*, the emperor of Gour attacked *Simul Nagar*. King Haripal, the ruler of *Simul Nagar*, left his kingdom and went to Basdinga Garh with the family by boat.

Paribar laiẏa Raja Nauka Asila Chorhe.
Praṇ Laiẏa Palailo Basḍiṅga Gaṛhe

[Meaning: The king came with the family to the boat. He fled to Basdinga Garh to save his life.]

Basdinga Garh, also known as Basuri, is located a few kilometres to the west of *Simul Nagar*. A river was situated between the *Basudinga Garh* and *Simul Nagar*. At present, *Basudinga Garh* is known as Basuri and the Kana Damodar is present in the abandoned form. Moreover, it indicates that Kanda Damodar was a large river in the medieval period.

In this chapter, the demarcation of PLC has been done using literary evidence and historical mapping. Although the literary evidence is indicative, narrative, and subjective in nature, it may be influenced by the author or the socio-cultural conditions in which it has been written. However, in this research, the location of the medieval period mentions the *Manasa Mangal* and *Dharma Mangal Kavya* signifying the presence of PLC of the Damodar River. Besides, the sequential disorder of a few locations of literary records has been found. Apart from that, the historical maps have a spatial dimension; therefore, it indicates the ancient river courses in a better way. The historical maps also allow us to detect the sequential changes in the Damodar River to confirm the presence of PLC.

3.4 Identification of Palaeochannel Using Geological Evidence

PLC represent the palaeo-fluvial condition in which they formed. The riverine system has high kinetic energy to flow the coarse materials. Therefore, PLC are associated with coarse materials such as well-sorted and rounded *kankar*, pebbles, and gravels with sand and clay alterations (Cao et al., 2019; Mahammad & Islam, 2021;

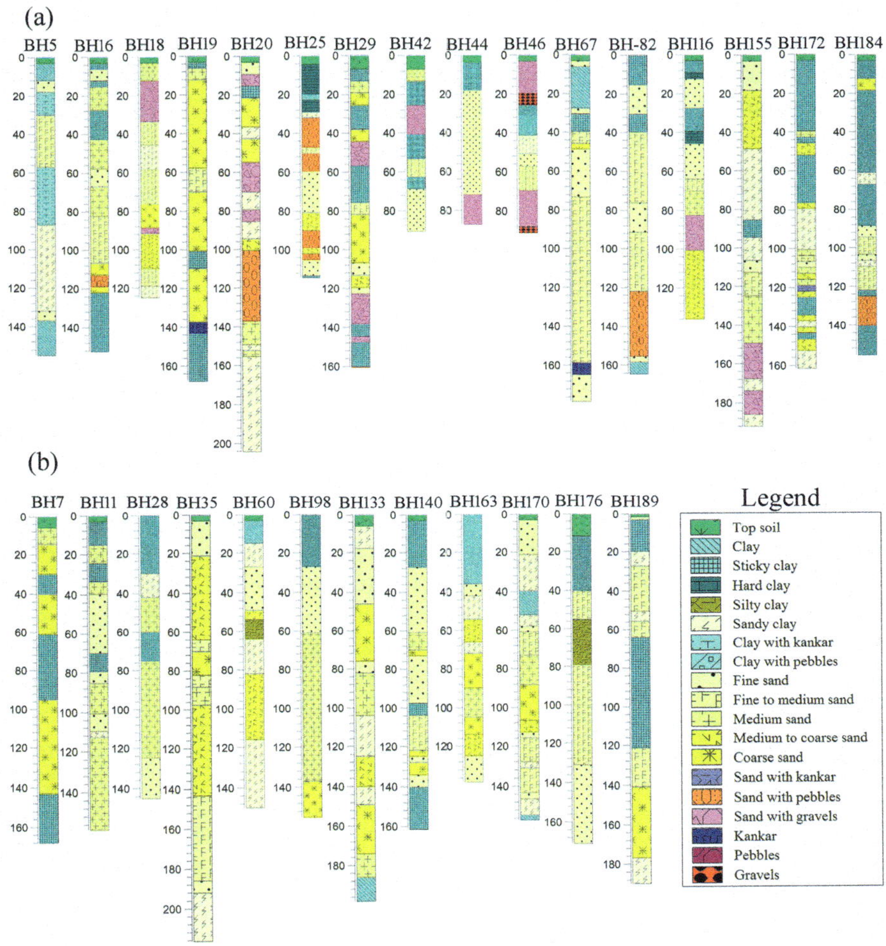

Fig. 3.4 Strip logs show the sub-surface lithological compositions of (**a**) PLC and (**b**) non-PLC

Bhadra et al., 2022). The demarcation of the PLC of the DFD has been executed with the help of 190 lithological logs collected from the Central Ground Water Board (CGWB), Government of India, State Water Investigation Directorate (SWID), Physical Health Engendering (PHE), and Agri-Irrigation department, Government of West Bengal. The borehole comprises different layers and portrays its characteristics. In this research, lithological logs of PLC (Fig. 3.4a) and non-PLC (Fig. 3.4b) have been separated. For instance, 16 logs from PLC and 12 logs from non-PLC locations have been represented. The lithological logs comprise sand and clay with coarse materials, such as *kankar*, pebbles, and gravel. In the borehole (BH5, BH16, BH18, BH19, BH20, BH25, BH29, BH42, BH44, BH46, BH67, BH82, BH116, BH115, BH172, and BH184 comprises sand mixed with gravels, pebbles and *kankar* in different depths. In addition, the non-PLC boreholes like

BH7, BH11, BH28, BH35, BH60, BH98, BH113, BH140, BH163, BH170, BH176, and BH189 contain sand, silt, and clay without the coarse materials. Thus, the strip logs with coarse materials characterize the palaeo riverine environment, indicating a typical fluvial architecture detailing the peculiarities as discussed implies the occurrence of palaeovalley in the study area.

3.5 Identification of Palaeochannel Using Geospatial Techniques

Identification of PLC using remote sensing data is based on surface and near-surface configuration of the PLC on satelliete images. According to CGWB (2016), PLC can be recognized based on some surficial appearance. PLC is confined in the wide valley with or without a narrow stream, which is unable to maintain its flow and fill up a significant part of the valley cross-section due to an inadequate supply of upstream water. PLC is associated with a series of waterbodies like ponds and lake that are dried up due to evaporation. Besides PLC can be found in buried conditions and they can be found based on the soil texture and moisture, linear vegetation cover, sediment characteristics, and aquifer conditions. Therefore, keeping in mind the PLC of the study area have been identified based on remote sensing datasets.

3.5.1 Datasets

Using RS datasets like Sentinel 1 (S1) A, Landsat 8 (L8) Operational Land Imager (OLI), and elevation profiles taken from Google Earth pictures, the PLC of the DFD have been identified. In a similar vein, field cross-sections and borehole data have been used to validate the likely PLC. For this study, two sets of L8 images with a 30 m spatial resolution (path/row 138/44 and 139/44, respectively, acquired on 18 December 2014 and 25 December 2014) were downloaded from the https://glovis.usgs.gov/. Since there are often fewer clouds in the sky and less moisture in the soil during the winter, the L8 datasets were gathered during this time. Additionally, the Alaska Satellite Facility (ASF) Vertex (https://search.asf.alaska.edu/#/) provided the S1A Interferometric Wide (IW) swath level 1 GRD data (path/frame 48/516 dated 11 November 2014) that was used for this investigation. Additionally, to illustrate the elevation zone of the study region, four sets of SRTM DEM (30 m) dated 23 September 2014 were acquired from https://earthexplorer.usgs.gov/. In addition, cross-sectional profiles across the PLC were created using elevation data from Google Earth Pro applications. In addition, the surface configurations of the PLC were depicted using the cross-sections of the likely PLC that were measured from the field. The PLC were validated using lithology data from the Central Groundwater Board (CGWB), Kolkata.

3.5.2 Methods

A strong methodological design that encompasses data collection, processing, and validations has been used to carry out the study. To put it another way, this study is divided into three main stages: preprocessing the RS data, identifying PLC by image processing methods, and validating the PLC that are identified using gathered geophysical data and field verification.

Preprocessing works to reduce the distortion that the sensor, atmosphere, and topography generate in satellite photos (Young et al., 2017). Using 'Top Atmospheric Correction', which transforms the digital number (DN) value to spectral reflectance in the QGIS 3.8.1 program to eliminate noise, the L8 data were preprocessed. Preprocessed bands 7 (SWIR 2), 5 (NIR), and 3 (Green) in the red, green, and blue (RGB) channels, respectively, were used to create the composite band because the combination of 7, 5, and 3 offers more atmospheric removal while maintaining natural colour (Butler, 2013). Using ArcGIS 10.4.1, the mosaic image was created using preprocessed information.

The S1 IW GRD data is not terrain corrected and is multi-looked at level 1.0. Consequently, preparing the S1 data is necessary for this investigation and was done so by following certain steps with the ESA SNAP (7.0) software. Initially, the sentinel data's geolocation quality was maximized by using the 'Apply Orbit File' feature. By normalizing the SAR backscatter signal throughout the whole S1 scene, the 'S-1 Thermal Noise Removal' was utilized to lessen the noise impacts in the inter-sub swath texture. Furthermore, for multi-swath acquisition mode software, this approach lessens the discontinuities between sub-swaths for the scene (Filipponi, 2019). Afterwards, low-intensity noise and incorrect data were reduced using the 'S-1 Remove GRD Border Noise' technique. The digital pixel value can also be converted into radiometrically corrected SAR backscatters by using the calibration step. Since it can maintain edges, line features, point targets, and texture information, the 'Refined Lee' speckle filtering was used for this investigation (Lee et al., 1994). In order to geocode the dataset and depict the SAR pictures geometrically, the 'Range-Doppler Terrain Correction' was implemented using SRTM DEM 1 arc second with the UTM/WGS84 coordinate system (ESA, 2020). Ultimately, the rectified backscatter data were transformed into the ArcGIS software's dB format.

Normalized Difference Water Index: The goal of this work is to identify open waterbodies and sediments that are rich in water by applying Gao's (1996) Normalized Difference Water Index (NDWI) approach to remotely sensed data. Using the raster calculator tool in ArcGIS based on Eq. 3.1, reflectance (ρ) of the near-infrared band (NIR) band (b5) and short-wave infrared band (SWIR) band (b7) of L8 OLI have been used to generate NDWI for this study. The NDWI value varies from +1 (indicating the existence of a marsh, moist soil, and waterbody) to −1 (indicating dry land surface or vegetation) (Lillesand et al., 2015).

$$\text{NDWI} = \rho\text{NIR} - \rho\text{SWIR} \,/\, \rho\text{NIR} + \rho\text{SWIR} \tag{3.1}$$

Since the plant cover on the PLC surface is well-grown, the linear pattern of the vegetation cover serves as a useful indicator of the presence of the PLC. Using the raster calculator tool of the ArcGIS software and Eq. 3.2, the Normalized Difference Vegetation Index (NDVI) for this study was derived in accordance with Tucker and Sellers (1986), taking into account the red (R) band (b4) and NIR band (b5) of Landsat 8 OLI. The vegetative index (NDVI) goes from +1 (vegetation) to −1 (waterbodies).

$$\text{NDVI} = \rho\text{NIR} - \rho R / \rho\text{NIR} + \rho R \tag{3.2}$$

The uncorrelated data are compressed into a narrower band using principal component analysis (PCA) (Rajani & Rajawat, 2011). The ArcGIS software was utilized to do PCA using the following bands: blue (b2), green (b3), red (b4), NIR (b5), SWIR 1 (b6), and SWIR 2 (b7). The resulting PCs were then utilized to delineate the PLC. Maćkiewicz and Ratajczak (1993) and Jackson and Edward (1991) state that Eq. 3.3 can be used to calculate the PCA as a covariance matrix.

$$y\begin{bmatrix} y1,1 & y1,2 & y1,n \\ y2,1 & y2,2 & y2,n \\ \cdots & \cdots & \cdots \\ ym,1 & ym,2 & yn,m \end{bmatrix} \tag{3.3}$$

where y 1, 1 is the variance of ith variable, x_i, and y_{ij} is the covariance between ith and jth variables.

3.5.3 Results

Using optical and microwave satellite datasets, the appearance of the PLC on and near the surface can be derived. The image processing methods listed below constitute the foundation for the defined PLC in the study area.

Since this band combination offers the most atmospheric penetration with true colour, the preprocessed bands 7, 5, and 3 of the L8 in RGB channels were used to create the fake colour composite (FCC) image for this investigation (Fig. 3.5a). The PLC are visually interpreted using the composite image to show the green and blue hues of the waterbodies and vegetation cover, respectively. In this study, the linear alignments of the vegetation cover and waterbody in the composite scene of L8 data represent the appearance of the PLC.

Information on near-surface characteristics that cannot be accurately visualized through the interpretation of optical satellite pictures is provided by the S1A GRD image. The backscatters of the S1 data in this investigation represent the moisture-rich sediment. Five classes have been identified from the processed S1 data backscatter in dB format. Radar scans show the water-rich layers as black tones, whereas the dry sediments are depicted as relatively lighter tones in the radar scene. As a

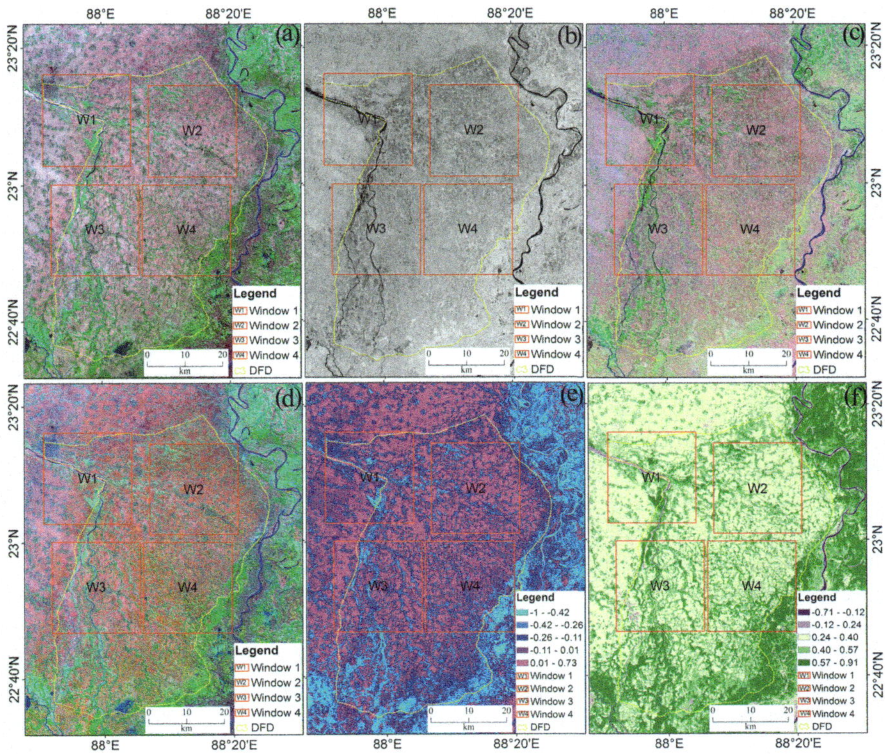

Fig. 3.5 PLC delineation using (**a**) false colour composite, (**b**) Sentinel 1A SAR, (**c**) fused image (L8 band 7, 5, 3 with Sentinel Sigma° VH dB format), (**d**) PCA, (**e**) NDWI, and (**f**) NDVI; the red box shows the window taken for evaluation of the PLC delineation efficiency (Source: Mahammad & Islam, 2021)

result, the class in the study identified as PLC ranges from 0.12 to 0.5 with linear alignments (Fig. 3.5b).

The Gaussian fusion technique was used to perform the multi-sensor fusion for this investigation from L8 optical and S1A GRD data, thereby improving the information included in both datasets. The output of the fusion data of L8 OLI and S1A GRD shows the direction of the PLC as green patches extending from the current research area's peak (Fig. 3.5c).

Moreover, three PCs—PC1, PC2, and PC3 in RGB—were used to create a composite PCA map. The red channel, which represents the colour blue, uses PC1. In the green channel, PC2 is utilized, whereas in the blue channel, PC3 is used. Water has enhanced the PLC. As a result, the spectral signature's reflectance differs greatly from that of the nearby objects. PCA has been utilized by Rajani and Rajawat (2011) to identify PLC. Principal components 1, 2, and 3 were employed as RGB. Wang et al. (2012) have identified the PLC using PCA combined with agricultural masking. The PLC in this study are connected to waterbodies and have a blue, linear configuration (Fig. 3.5d). According to the PCA result, component 1 accounts for

Table 3.1 Variance and eigenvalues derived through the PCA

Indicators	Component 1	Component 2	Component 3
% Variance	81.41	16.68	1.92
Eigenvalues	0.00835	0.00171	0.00020

81.41% of the variation overall, with components 2 and 3 following closely behind (16.68% and 1.92%) (Table 3.1).

The NDWI value in the studied area falls between −1 and 0.73. Using natural breaks, the image has been categorized into five classes. The NDWI values have been used to classify the first class (− 1 to −0.42), second class (−0.42 to −0.26), third class (−0.26 to −0.11), fourth class (−0.11 to −0.01), and fifth class (−0.01 to 0.73). The suggested PLC are the blue linear segments connected to waterbodies, with values ranging from −1.4 to 0.42 (Fig. 3.5e).

Likewise, the highest intensity of green vegetation cover is indicated by a positive NDVI value. Conversely, the negative number indicates the lowest level of vegetation cover intensity and is represented in blue. The NDVI value in the studied area falls between −0.71 and 0.91. Using natural breaks, the image has been categorized into five classes. Based on the NDVI values, the following classes have been identified: first class (−0.71 to −0.12), second class (−0.12 to 0.24), third class (0.24 to 0.40), fourth class (0.40 to 0.57), and fifth class (0.57 to 0.91). The study suggested PLC are the green linear segments of the high NDVI value (0.57 to 0.91) (Fig. 3.5f).

The paths of six major PLC are identified based on image processing algorithms that combine observations from topography maps and Google Earth pictures with both microwave and optical datasets (Fig. 3.6). Aside from that, lots of marshes and river cuts have been found. The following has been explored regarding the alignment and dispersion of the PLC.

PLC I, which originated in Belna and split off close to Saktigarh, has been identified by radar and optical data. The northern portion of the PLC runs northward to Hatgobindapur, and then it flows eastward through Satgachhia, Jabui, and Nabastha to Singail. PLC I's southern portion runs eastward to join PLC II close to Gopinathpur.

After splitting off from the Damodar River close to Palla Colony in Barddhaman, PLC II runs in a meandering or sinuous manner for a few kilometres northeast. It splits after Gopinathpur, with the upper section known as the Behula Nadi flowing in a meandering manner towards the east, passing through Amadpur, Parhati, Debpur, Magra, Hatbaksa, Tantibaksa, and Sitarambati before arriving at Kutubpur. Following Kutubpur, the PLC bends right and travels south to Baidyapur. It then flows east via Mirhat and Senerdanga before joining the Hooghly River close to Kalna. From there, another portion of the PLC runs south to Somra. Palaeochannel II, a section of which originates in Debpur, passes through Shridharpur, Barwa, and Isabpur before joining the Behula Nadi at Senerdanga. The lower portion of the PLC II flows eastward from Gopinathpur through Memari, Bahabpur, and Kabirpur, ultimately reaching Dhamas.

The Chanchai is the source of PLC III, which is likely the remnants of the ancient Damodar River. It passes through Ajhapur, Chokhanda, Berela, Boinchi, Simlagarh

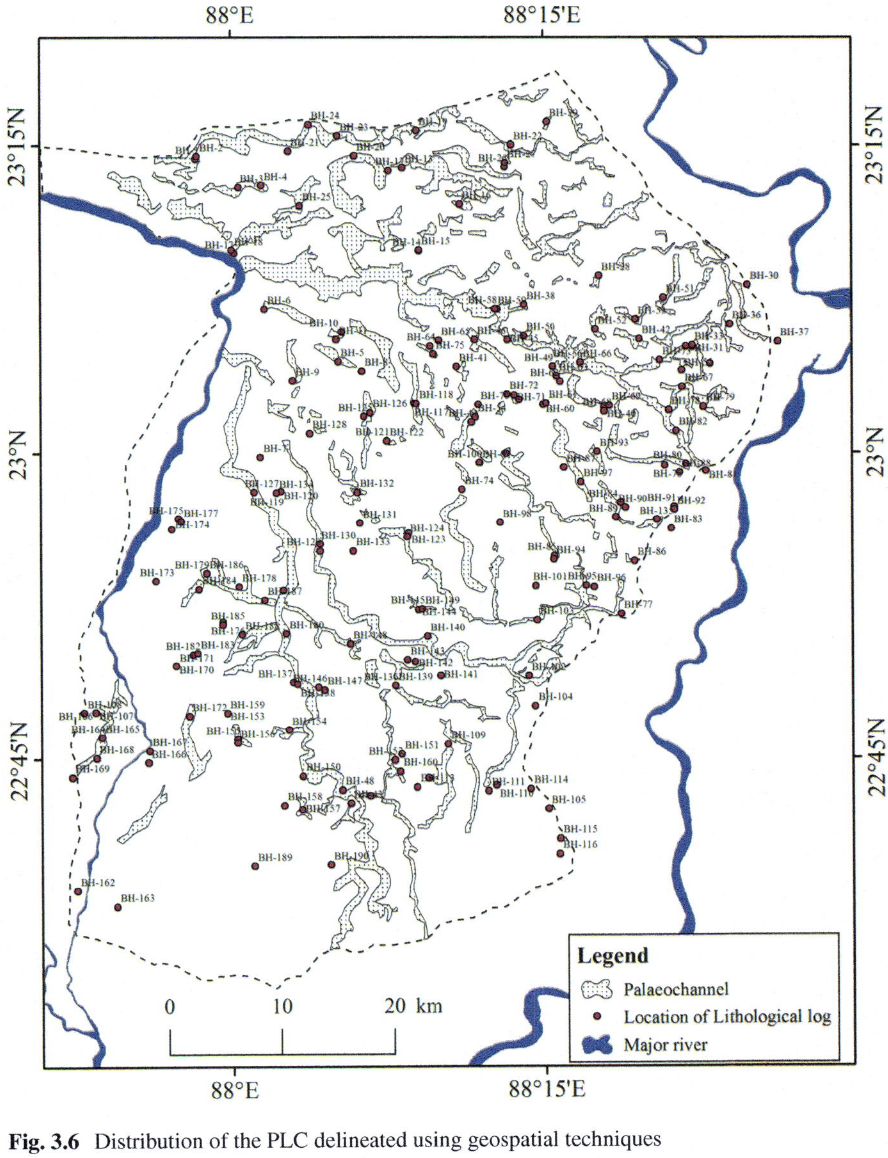

Fig. 3.6 Distribution of the PLC delineated using geospatial techniques

Pandua, Khanyan, and Chanparai. In the studied region, PLC III is separated numerous times and forms a dendritic pattern that flows eastward. The PLC splits off close to Debipur and flows through Tola, Jamna, Bara Baharkuli, and finally the Behula Nadi at Natagarh in a northeastern direction. At Berela village, another portion splits off from the PLC and heads south and southeast via Ramnagar, Dwarbasini, and Khamarpara. It also splits close to Balikukhari, with the split channel flowing southward. Near Pandua town, PLC III is further divided into two branches: one flows via Mondolai and Depara in a northeastern direction all the way to Masra, while the other through Namajgram, Mandaran, and Dadpur in a southeasterly direction all the way to Magra.

Based on satellite data, PLC IV has been identified, with its upper reach separated into two sections. The PLC's northernmost portion is limited to the Kantal Nadi; it begins close to Masagram and passes through Nabagram, Bhastara, Kamrai, Aima, and Hodla. The Kantal Nadi is referred to as Ghea Nadi after Hodla. The Mahindar Nadi originates in Mahindar and runs through Chopa and Jaugram before meeting the Ghea Nadi close to Beraberi. It also flows through Hatkamalpur, Dhaniakhali, Talbona, Dhanaitikari, Nalitajol, and Laksmanpur in the southern portion.

The current Kunti or Kana River, which splits from the Damodar River at Selimabad and flows in a semi-circular pattern southward, eastward, and finally northeast, is identified as PLC V. This channel was also identified using the aforementioned techniques. It eventually empties into the Hooghly River at Noaserai.

PLC VI, or the Kana Damodar, splits off from the Damodar River at Selimabad and flows almost entirely south through the following locations: Dighir, Ballavipur, Baligari, Dwipa, Jangipara, Prasadpur, and Jagatballavpur. Eventually, it converges at the Hooghly River's outfall point, which is close to Uluberia. At Baligari, the PLC splits into two, with a portion flowing southeast and then south via Kaikala, Haripal, Sipai Gacchi, Furfura, Hantal, and Gondalpara. In addition, numerous sections of the PLC have been discovered as linearly aligned waterbodies, abandoned channels, and river cut-offs. In addition, a number of back wetlands have been located using the previously indicated image processing methods.

Nonetheless, six significant PLC have been identified that resemble the PLC distinguished by Niyogi (1975) and Mallick and Niyogi (1972). The prior investigation revealed PLC, which are old, deserted canals of various sizes that are primarily restricted to the study area's northern region. In addition, the PLC in the current research area differs from those in earlier studies in terms of length and width. The surface profile generated from Google Earth shows that the width of PLC I and II is lower than that of the PLC shown in the prior research, which validates the width of the PLC for this study. While PLC IV has been partially identified as a portion of the Kantal Nadi in the earlier study, it has been identified as Ghia Nadi in this investigation, which has a longer length. The Kana-Kunti Nadi and Kana Damodar Rivers are identified as PLC V and VI in this study, whereas they were previously defined as minor natural levees in the prior study.

3.5.4 *Discussion*

3.5.4.1 Efficiency of the Image Processing Techniques

Using the four research area windows that were chosen, a comparison of image processing methods that use both radar and optical data to identify PLC has been carried out. This will demonstrate how effective the image processing methods are.

All of the PLC off-take stations from the Damodar River are included in window 1, which is situated in the highest point of the DFD (Fig. 3.7a–f). Six locations throughout the PLC have been chosen in window 1 in order to evaluate the effectiveness of the optical and radar satellite data. The radar satellite data at Nabastha clearly shows the presence of the two PLC I segments in dark tones, which is barely discernible in the optical L8 data. The PLC is visible in the PCA image as well, and it appears more distinct than in the composite image. Additional information

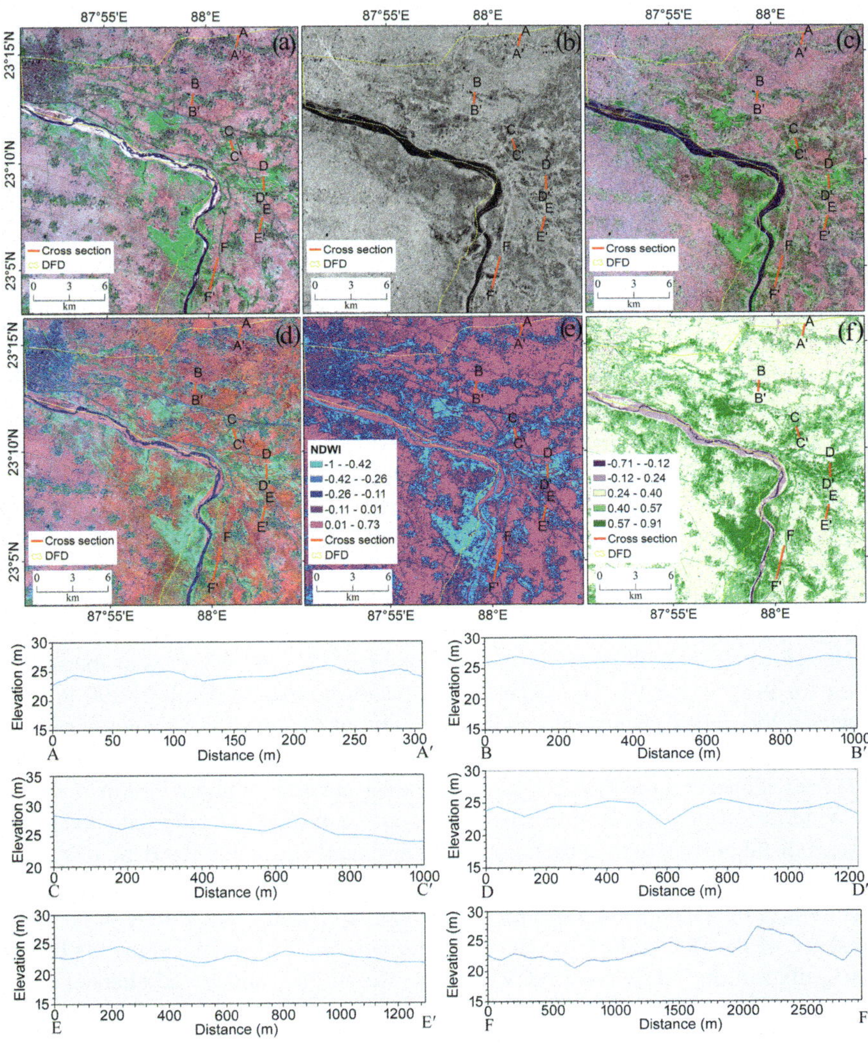

Fig. 3.7 PLC detection efficiency of image processing techniques for window 1: (**a**) false colour composite, (**b**) Sentinel 1A SAR, (**c**) fused image, (**d**) PCA, (**e**) NDWI, (**f**) NDVI; A-A′, B-B′, C-C′, D-D′, E-E′, and F-F′ depict the PLC locations within the cross-profiles (Source: Mahammad & Islam, 2021)

regarding the existence of the PLC is provided by the examination of the NDVI techniques, which indicates that the PLC's width is 420 m rather than 470 m as the NDVI technique suggests. Additionally, the transverse profile A-A′ close to Nabastha shows that the PLC is present and has a width of around 470 m at a depth of 2 m. The L8 data at Putunda village, next to Saktigarh railway station, shows the southern portion of PLC I as linear vegetative cover, whereas the radar satellite data detects the channel as having a darker tone than the surrounding area (Fig. 3.7a–f). The appearance of the PLC is more clearly visible in the PCA images. Furthermore, the PLC's width of 280 meters is seen by the NDWI study, but the PLC's width of 210 m is shown by the NDVI analysis. A transverse profile B-B′ has been drawn over the lower portion of PLC I at this position, indicating the presence of a PLC with a width of 410 m and a depth of 2 m. Optical and radar data have been used to compare in PLC II at Sahapur, Barddhaman (Fig. 3.7a–f). The research reveals that the PLC is represented by radar data to be wider than that of optical data. The PLC appears to be 307 m wide on the NDVI, while the NDWI analysis indicates that it is 310 m wide. A cross-section that illustrates the PLC's evidence of existence and its approximate 550-meter width and 2-m depth has been drawn. Whereas the optical data indicates the presence of the PLC in terms of the presence of waterbodies and vegetation cover, the radar data in PLC III, near Ajhapur, has demonstrated the PLC's width in a dark tone. It also shows that the radar data's detection of the PLC's width is greater than the optical data's observation of the same breadth. While the NDVI technique reveals a width of around 708 m, the NDWI analysis indicates the presence of a PLC with a width of approximately 680 m. The cross-section C-C′ indicates that the width of PLC III near Ajhapur is 1.24 km, with a depth of 3 m. PLC IV, located close to Nabagram, has a width of approximately 450 m, which is less than what the radar data indicates. The composite picture and PCA also show a width of about 550 meters. In addition, the NDWI and NDVI indicate that the PLC is 430 m wide. The PLC near Nabagram has a width of 540 m and a depth of 2 m, according to the cross-section D-D′. Comparably, measurements of the PLC's width at Halara, close to Selimabad, have also been made. This PLC serves as the source point for PLC V and VI. The radar data indicates a channel width of 290 meters; however, the optical data shows a channel width of 585 m. The channel's width is shown by the NDWI and NDVI to be 465 and 575 m, respectively. The cross-section at Halara shows that the PLC is 4 m deep and has a width of about 710 m.

Window 2 (Fig. 3.8a–f) shows the lower courses of the PLC and is situated in the eastern distal portion of the DFD. Based on the examination of radar data, it is noted that at Belara, which has a width of 430 meters, PLC III is plainly visible in a black tone. The optical data, which includes RGB and PCA, cannot accurately observe the presence of the PLC at this site. In a similar vein, the PLC's appearance is hardly discernible in the NDWI and NDVI output, which only shows the channel's few meters of width. Similar to the data output observations, the transverse profile G-G′ at Berala shows that the width of PLC III is 430 m at a depth of 9 m. The optical data, including RGB and PCA, is unable to accurately detect the presence of the PLC at the second location at Baidyapur, while it does show the breadth of PLC II as 650 m. In a similar vein, the PLC's appearance is hardly discernible in the NDWI

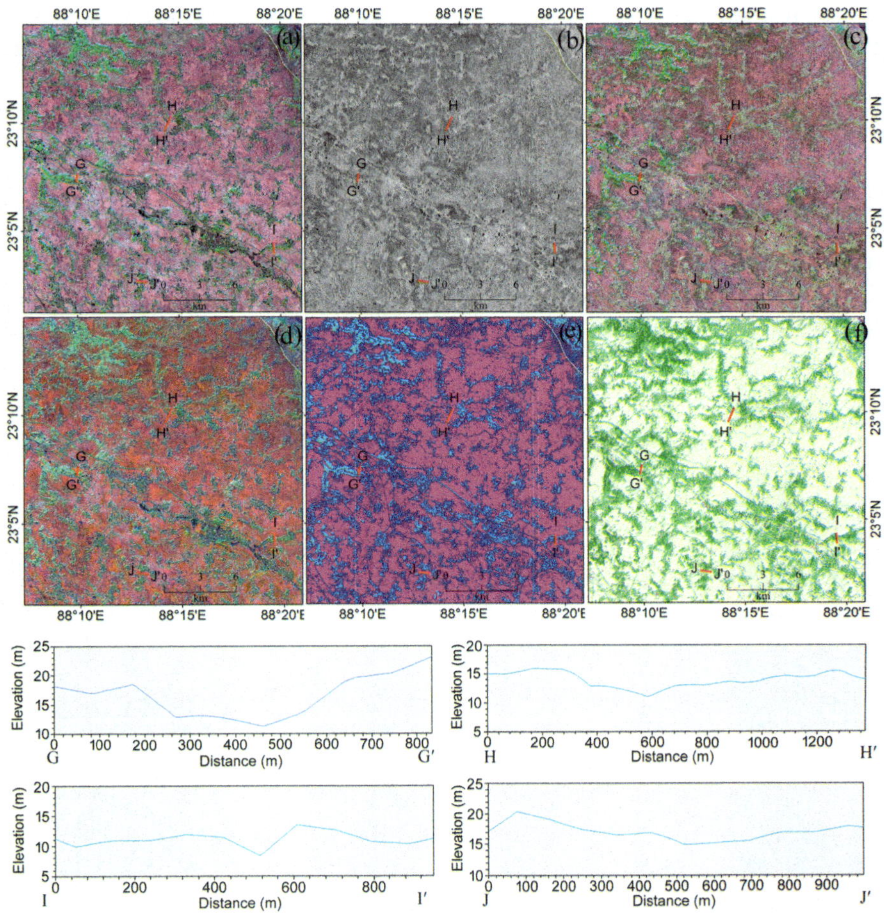

Fig. 3.8 PLC detection efficiency of image processing techniques for window 2: (**a**) false colour composite, (**b**) Sentinel 1A SAR, (**c**) fused image, (**d**) PCA, (**e**) NDWI, (**f**) NDVI; G-G′, H-H′, I-I′, and J-J′ depict the PLC locations within the cross-profiles (Source: Mahammad & Islam, 2021)

and NDVI output, which only shows the channel's few meters of width. The PLC near Baidyapur has a width of 810 m and a depth of 4 m, according to the cross-section H-H′. The radar data at Chapatta indicates that PLC III is 650 meters wide; however, the optical data shows that it is 580 meters wide. The widths of 495 and 310 m are represented by the NDVI and NDWI, respectively.

PLC III at Chanpta has a width of 320 m and a depth of 3 m, according to the cross-section I-I′. The S1A data has shown that PLC III near Gopalnagar is 750 m wide. The optical L8 data, which includes RGB and PCA, cannot clearly observe the presence of the PLC at this site. In the same way, the PLC's representation in the NDVI and NDWI output is similarly scarcely discernible. The cross-section J-J′ reveals that the PLC is about 710 m wide and 2 m deep.

The restricted courses of the PLC are depicted in window 3, which is situated in the more recent Tarakeshwar Fan Delta (Fig. 3.9a–f). The four locations in the current window are meant to serve as reference points for the image processing methods used in this investigation. First, the examination of radar data near the Bhangamora indicates that the original Damodar channel was wider than the current course by 670 m. The optical data, which includes RGB and PCA, can be used to directly view the current water course and, by using the vegetation cover along the banks, to suggest the wider ancient course. Additionally, the NDVI and NDWI show that the river's course is wider. A 900 m shallow depression across the river is revealed by measuring the transverse profile K-K′ close to Bhangamora, indicating the presence of PLC. The S1A data at the second location at Dighir clearly shows

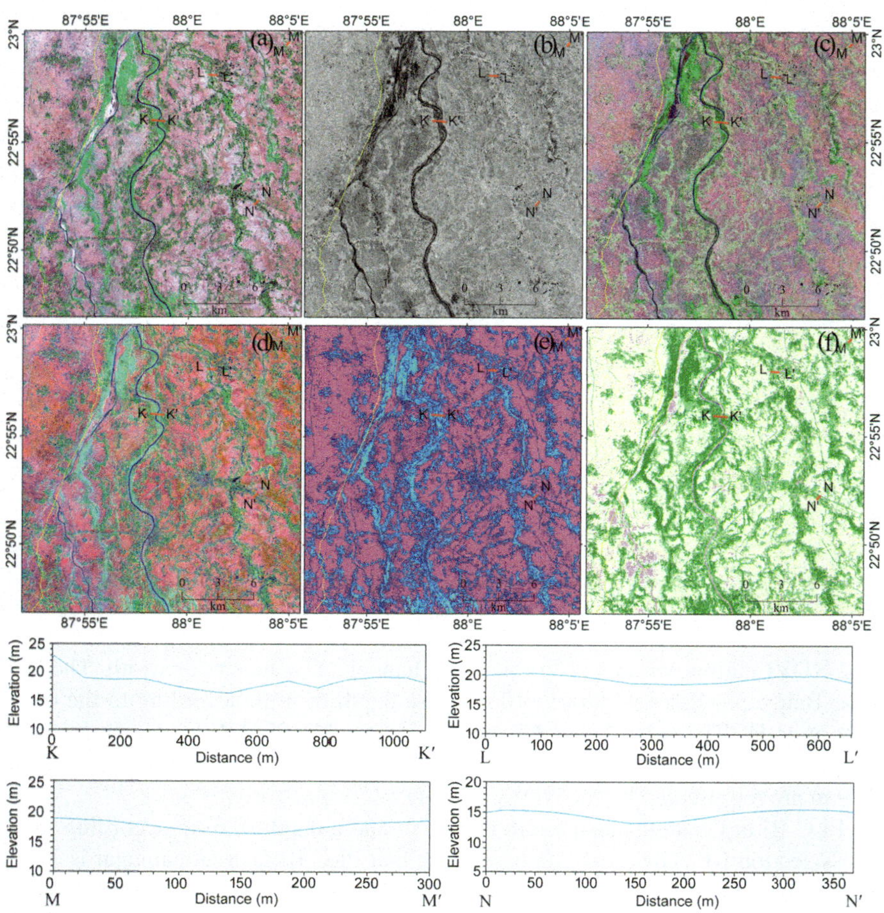

Fig. 3.9 PLC detection efficiency of image processing techniques for window 3: (**a**) false colour composite, (**b**) Sentinel 1A SAR, (**c**) fused image, (**d**) PCA, (**e**) NDWI, (**f**) NDVI; H-H′, I-I′, and J-J′ depict the PLC locations within the cross-profiles (Source: Mahammad & Islam, 2021)

the breadth of PLC VI, whereas the optical data, which include RGB and PCA, depict PLC as being surrounded by a linear vegetative cover. Similarly, the NDWI and NDVI outputs show the PLC's appearance. With a depth of 2 m, the cross-section L-L′ indicates the PLC's width of around 475 m. The radar data at Hatkamalpur has revealed the width of PLC IV to be 180 m, a feature that is scarcely discernible in the optical data. Only a few meters' worth of breadth is represented by the NDVI and NDWI. According to the cross-section M-M′, PLC VI has a width of 240 m and a depth of 3 m. The S1A data has shown that the PLC near Bagbari is 100 m wide. The PLC with a width of 280 m is represented by the optical L8 data with RGB and PCA included. The PLC's width is shown to be 130 m by the NDWI and 125 m by the NDVI, respectively. The PLC is approximately 295 m in width and 2 m in depth, as indicated by the cross-section N-N′.

The lower courses of the PLC are depicted in window 4, which is situated in the southeast of the DFD (Fig. 3.10a–f). The four sites—windows 1–4—have been chosen so that the image processing methods used in this study can be compared. Firstly, the study of radar data near Bandipur indicates that the PLC V has a diameter of 113 m. The L8 data indicates that the PLC is 150 m wide. Additionally, the PLC's width is shown by the NDWI and NDVI to be about 60 and 150 m, respectively. The 230 m shallow depression across the river, as revealed by the measurement of the transverse profile O-O′, suggests the presence of PLC. The S1A data at the second location, Talbona, indicates that the width of PLC IV is around 350 m, whereas the optical data, which includes RGB and PCA, indicates that the breadth of the PLC is approximately 105 m. Comparably, the PLC's width is shown to be approximately 155 m in the NDWI and 180 m in the NDVI result. The PLC measures 330 meters in width and 2 m in depth, according to the cross-section P-P′. PLC III's width near Dwarbasini has been determined by radar data to be 310 m; however, optical evidence indicates that the PLC is only 135 m wide. According to the NDWI and NDVI, the breadth is 100 and 138 m, respectively. The PLC measures 330 m in width and 5 m in depth, according to the cross-section Q-Q′. The radar data has recorded a width of 295 m for the PLC IV near Nalitajol; however, the optical L8 data, which includes RGB and PCA, indicates a width of 100 m. The PLC's width is shown to be 105 meters in the NDWI output and 115 meters in the NDVI output. The cross-section R-R′ displays the PLC's width of about 330 meters and its depth of 2 m.

3.5.4.2 Evidence from the Surface Cross-Sections

Based on the Dumpy Level Study, cross-sections through the PLC at Simalagarh and Boinchi have been drawn. The cross-section at the first site (Boinchi) is 119 m long (88°12′05.8″E to 88°12′09.6″ E and from 23°07′09.6″ N to 23°07′07.5″ N) with an elevation range of 9.58 to 11.68 m. After 53 m, the height gradually decreases until it reaches 101 m, at which point it rises again. As a result, a valley-like depression that appears in the cross-section is thought to be a PLC in this region (Fig. 3.11a). In a similar vein, the second site's (Simlagarh) cross-section has a

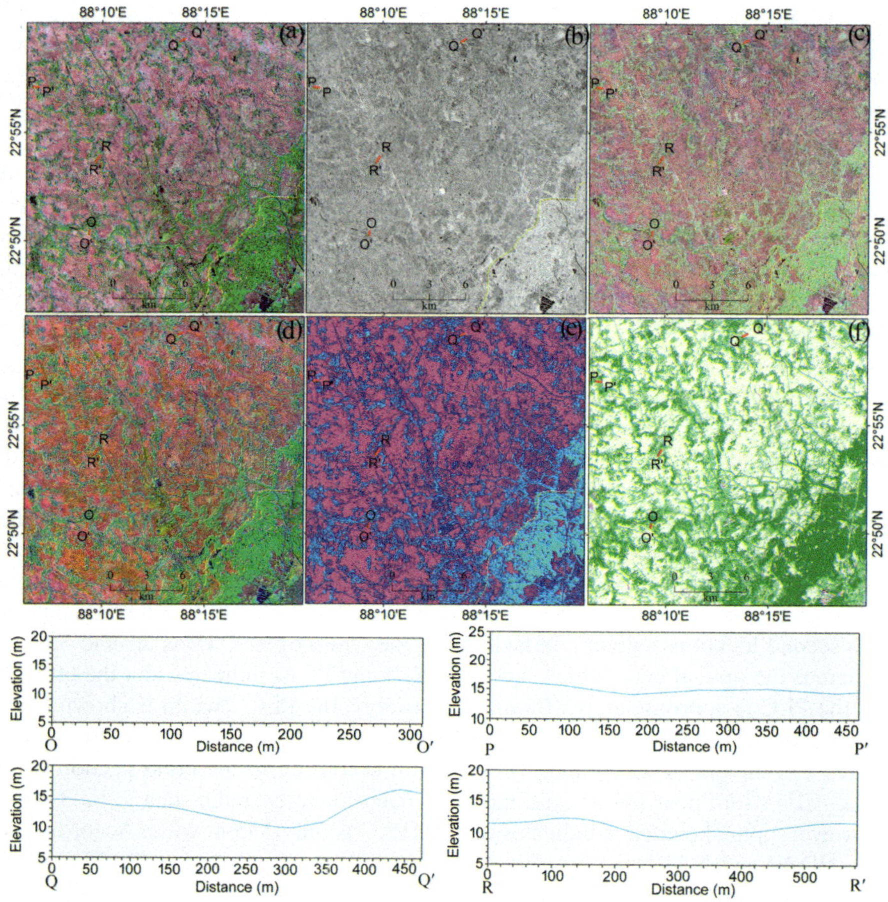

Fig. 3.10 PLC detection efficiency of image processing techniques for window 4: (**a**) false colour composite, (**b**) Sentinel 1A SAR, (**c**) fused image, (**d**) PCA, (**e**) NDWI, (**f**) NDVI; K-K′, L-L′, M-M′, and N-N′ depict the PLC locations within the cross-profiles (Source: Mahammad & Islam, 2021)

length of 157.8 meters (88°14′29.7″ E to 88°14′29.3″ E and from 23°05′09.0″N to 23°05′14.4″ N). The difference in elevation is 10.75–12 m. There is a small indentation in the cross-section in the segment that runs from 39.8 to 120 m. Moreover, this is identified as a possible PLC (Fig. 3.11b) in this area. Furthermore, it is easy to distinguish the possible PLC from the Google Earth image, which appears as continuous ponds and waterbodies in a straight pattern close to Boinchi and Simalagarh (Fig. 3.12a). The observation of elongated channel depressions during the field

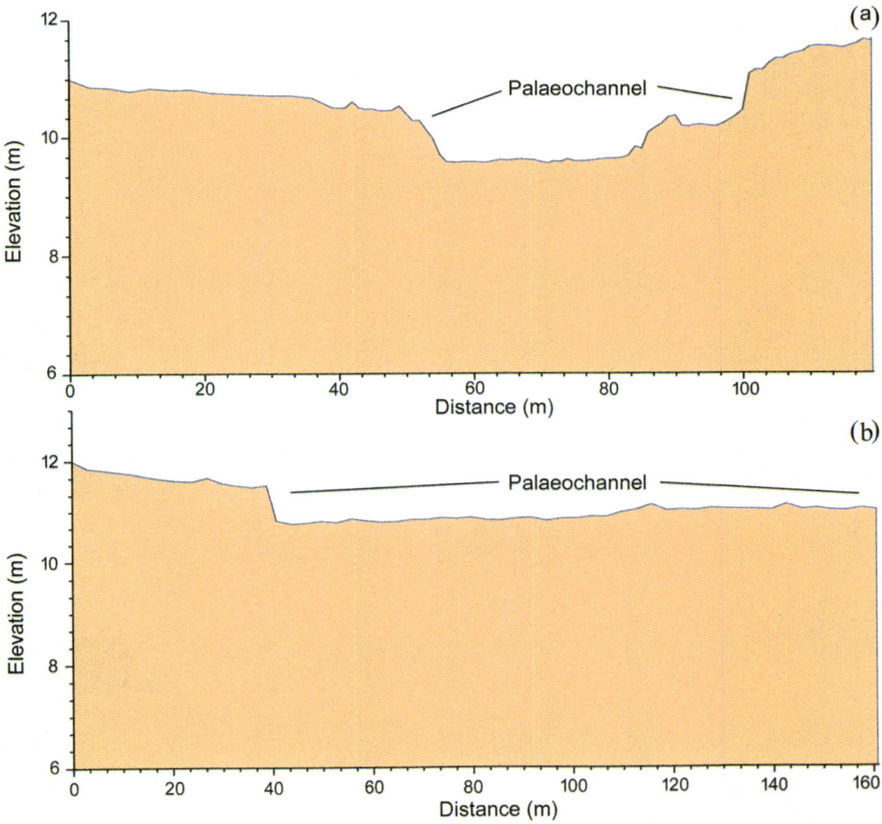

Fig. 3.11 PLC existence at the depressed part of the cross-section at (**a**) Boinch and (**b**) Simlagarh (Source: Mahammad & Islam, 2021)

survey is notable. A portion of Simlagarh's PLC is utilized as a local waterbody and is found to be in ponding conditions (Fig. 3.12b) and at some point, fresh sediments fill the PLC, turning them into an agricultural field (Fig. 3.12c, d). Furthermore, the field visits demonstrate that groundwater is extracted for home and agricultural purposes from the area thought to be a PLC field of agriculture (Fig. 3.12e). However, it is not viable to determine PLC only by surface expression. Only their likely locations may be inferred from this evidence. Thus, to verify their positions, the subsurface view is crucial.

Fig. 3.12 PLC appearance: (**a**) PLC surface expression in Pandua and surrounding area on Google Earth image, (**b**) PLC ponding at Simlagarh, (**c**, **d**) agricultural practice on PLC at Boinchigram and Jabui village near Memari, and (**e**) groundwater withdrawal near PLC at Jabui village (Source: Mahammad & Islam, 2021)

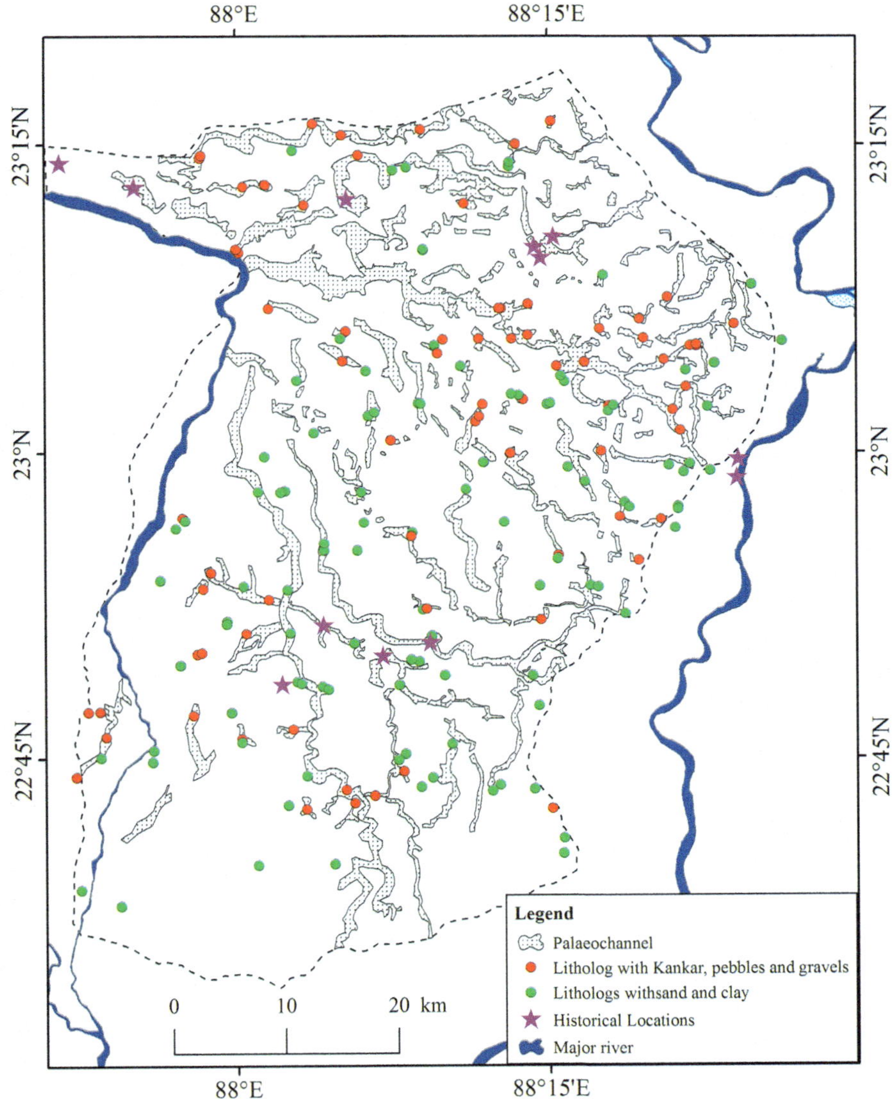

Fig. 3.13 Integration of delineated PLC with the literary evidence, historical records, and lithological characteristics

3.6 Critical Analysis of PLC Detection Methods

Several PLC identification techniques, such as literary evidence, historical mapping, remote sensing, and geophysical data analysis, are used in this chapter. PLC detection is directly related to the approaches' integration (Fig. 3.13). The found PLC resemble the current rivers in the research area's abandoned state. The effectiveness analysis shows that while the SAR data depicts the study's near-surface properties, the optical data offers more precise information about surface configurations. Additionally, the surface profile has been used to quantify the breadth of PLC, indicating the quality of the radar and optical data. The study's validation showed that the data and techniques employed were suitable for identifying the PLC and comprehending the historical fluvial system with justifications. Furthermore, lithological strata show that the mix of coarse materials like kankar, pebbles, and gravels with clay and sand is linked to PLC in the research area. This suggests the kinetic energy of the riverine system. Remote sensing data indicates the presence of coarse material along the delimited PLC, hence confirming the existence of PLC. Additionally, the PLC that were defined were dispersed radially from the highest point of the DFD, indicating the process of delta development linked to variations in sea level.

PLC have been identified, thanks to the integration of historical maps, literary records, optical and radar data, and geophysical data. This integration also highlights the important relationship between the drainage courses in the studied area's history and present. Various approaches were used in this study to identify the six major PLC. This study established the presence of the Damodar River's ancient route, which served as the river's primary course throughout a variety of historical periods. This study also shows that the rapid alteration in the Damodar River's lower course gave rise to the PLC. The historic Damodar River's route is also revealed by the delineated PLC, and this information may be crucial for managing groundwater in the semi-critical blocks of the Damodar Fan Delta. Furthermore, because remote sensing images cannot identify features hidden deeper beneath the surface layer, it is challenging to identify the deeply buried portion of the PLC. As a result, future research may concentrate on applying intensive geophysical and geoelectrical surveys in conjunction with cutting-edge geospatial techniques.

Chap. 4 will focus on the availability and the quality of the surface and groundwater in the DFD to detect the need for palaeo channels in water resource management.

References

Bandyopadhyay, S. (1996). Location of the Adi Ganga palaeochannel, South 24 Parganas, West Bengal: A review. *Geographical Review of India, 58*(2), 93–109.

Bhadra, B. K., Jain, A. K., Dutta, A., Meena, H., Rehpade, S. B., & Rao, S. S. (2022). Multi-sensor satellite images and resistivity survey for the evolution of Sutlej PLC and groundwater aquifer potential in Punjab, Northwest India. Geocarto International, 1–21.

Bhattacharyya, K. (2011). *The lower Damodar River, India: Understanding the human role in changing fluvial environment*. Springer.

Biswas, K. R. (2001a). *Rivers of Bengal* (Vol. I). Government of West Bengal.

Biswas, K. R. (2001b). *Rivers of Bengal* (Vol. II). Government of West. plate-9(1), 10 and 11.

Butler, K. (2013). ArcGIS Blog. Retrieved October 19, 2020, from ESRI: https://www.esri.com/arcgis-blog/products/product/imagery/bandcombinations-for-landsat-8/

Cao, Y., Cao, G., & Shan, W. (2019). Morphological characteristics of the paleochannel of the Yihe River in the Last Glacial Maximum, Shandong Province, China. *Physical Geography, 40*(4), 307–322.

CGWB (2016). Palaeochannel of North West India: Review and assessment. Report of the expert committee to review available information on palaeochannel, 1–127.

Chakraborty, K., Mukherjee, I., & Ghosh, S. (2011). Historical and geo-environmental appraisal of changing growth centre, urbanization and behavioural dynamics of the Rivers of Barddhaman town, West Bengal. *International Journal of Current Research, 3*(11), 216–224.

Charlton, R. (2007). *Fundamentals of fluvial geomorphology*. Routledge.

ESA (2020). User Guides, Sentinel 1 SAR. Retrieved October 19, 2020, from ESA: https://sentinel.esa.int/web/sentinel/user-guides/sentinel-1-sar

Filipponi, F. (2019). Sentinel-1 GRD preprocessing workflow. In *International electronic conference on remote sensing* (p. 11). MDPI.

Gao, B. C. (1996). NDWI-A normalized difference water index for remote sensing of vegetation liquid water from space. *Remote Sensing of Environment, 58*(3), 257–266.

Ghosh, S. (2011). Hydrological changes and their impact on fluvial environment of the lower Damodar basin over a period of fifty years of damming the mighty Damodar River in Eastern India. *Procedia-Social and Behavioral Sciences, 19*, 511–519.

Ghosh, P. K., & Jana, N. C. (2019). Historical evidences in the identification of palaeochannels of Damodar river in western Ganga-Brahmaputra Delta. Quaternary Geomorphology in India: Case Studies from the Lower Ganga Basin, 127–138.

Ghosh, S., & Mistri, B. (2012). Hydrogeomorphic significance of sinuosity index in relation to river instability: A case study of Damodar River, West Bengal, India. *International Journal of Advances in Earth Sciences, 1*(2), 49–57.

Harris, D. R. (1980). Exhumed paleochannels in the Lower Cretaceous Cedar Mountain Formation near Green River, Utah. Ph.D. dissertation, Brigham Young University.

Jackson, J. E., & Edward, A. (1991). *User's guide to principal components* (p. 40). John Willey Sons..

Lee, J. S., Jurkevich, L., Dewaele, P., Wambacq, P., & Oosterlinck, A. (1994). Speckle filtering of synthetic aperture radar images: A review. *Remote Sensing Reviews, 8*(4), 313–340.

Lillesand, T., Kiefer, R. W., & Chipman, J. (2015). *Remote sensing and image interpretation*. John Wiley & Sons.

Maćkiewicz, A., & Ratajczak, W. (1993). Principal components analysis (PCA). *Computers & Geosciences, 19*(3), 303–342.

Mahammad, S., & Islam, A. (2021). Identification of palaeochannels using optical images and radar data: A study of the Damodar fan Delta, India. *Arabian Journal of Geosciences, 14*(17), 1702.

Mallick, S., & Niyogi, (1972). Application of geomorphology in groundwater prospecting in the Alluvial Plains around Burdwan, West Bengal.

Mohammed-Aslam, M., & Balasubramanian, A. (2010). History of river channel modifications—A review. *Journal of Ecology and the Natural Environment, 2*(10), 207–212.

Mukerjee, R. (1938). *The changing face of Bengal. A study in riverine economy*. Calcutta University Press, University of Calcutta.

Niyogi, D. (1975). Quaternary geology of the coastal plain of West Bengal and Orissa. *Indian Journal of Earth Science, 2*, 51–61.

Page, K., Nanson, G., & Price, D. (1996). Chronology of Murrumbidgee river PLC on the Riverine Plain, southeastern Australia. *Journal of Quaternary Science: Published for the Quaternary Research Association, 11*(4), 311–326.

Rajani, M. B., & Rajawat, A. S. (2011). Potential of satellite based sensors for studying distribution of archaeological sites along palaeochannels: Harappan sites a case study. *Journal of Archaeological Science, 38*(9), 2010–2016.

Rajawat, A. S., Verma, P. K., Nayak, S., Rajawat, A. S., Verma, P. K., & Nayak, S. (2003). Reconstruction of palaeodrainage network in Northwest India: Retrospect and prospects of remote sensing based studies. *Proceedings-Indian National Science Academy Part A, 69*(2), 217–230.

Rathore, V. S., Nathawat, M. S., & Champatiray, P. K. (2010). Palaeochannel detection and aquifer performance assessment in Mendha River catchment, Western India. *Journal of Hydrology, 395*(3–4), 216–225.

Rudra, K. (2010). Banglar Nadikatha. Sahitya Sansad.

Tucker, C. J., & Sellers, P. J. (1986). Satellite remote sensing of primary production. *International Journal of Remote Sensing, 7*(11), 1395–1416.

Wang, X., Guo, Z., Wu, L., Zhu, C., & He, H. (2012). Extraction of PLC information from remote sensing imagery in the east of Chaohu Lake, China. *Frontiers of Earth Science, 6*(1), 75–82.

Wray, R. A. (2009). Palaeochannels of the Namoi River floodplain, New South Wales, Australia: The use of multispectral Landsat imagery to highlight a late quaternary change in fluvial regime. *Australian Geographer, 40*(1), 29–49.

Young, N. E., Anderson, R. S., Chignell, S. M., Vorster, A. G., Lawrence, R., & Evangelista, P. H. (2017). A survival guide to Landsat preprocessing. *Ecology, 98*(4), 920–932.

Chapter 4
Surface Water and Groundwater Dynamics and Quality Assessment

4.1 Dynamics of Surface Waterbodies Using Artificial Intelligence and Machine Learning Algorithms

The term 'surface water dynamics' reflects the spatiotemporal changes in the surface waterbody. Surface waterbodies play a vital role in the development of groundwater resources (Arkoprovo et al., 2012; Krishnamurthy et al., 2000). Thus, inventory mapping and management of surface waterbodies are of vital significance. In the context of the DFD, the dynamics of the surface waterbodies are examined using some geo-statistical tools, machine learning applications and their proper validations. In this chapter, another important dimension of surface water resources is the evaluation of water quality for irrigation, drinking, and ecological health. Therefore, the purpose of this chapter is to map the dynamics of change in the waterbodies of the study area and evaluate their quality. These are efficient for getting a better picture of the water resources inventory that may be useful for balancing the surface and groundwater resources in the DFD and its future development. The following sections are committed to these ends.

Surface waterbodies are essential resources to maintain environmental balances and for human needs. Therefore, the extraction and mapping of surface waterbody have been conducted by many researchers using remote sensing data (Acharya et al., 2018; Bijeesh & Narasimhamurthy, 2020). For this purpose, the extraction of the surface waterbodies has been done using both the supervised image classification and band ratio method in several studies (Saputra & Lee, 2019; Feyisa et al., 2014). In this research, LULC image classification has been implemented for extraction of the surface waterbody and its spatiotemporal changes. Moreover, the intensity analysis has been conducted to trace the dynamics of LULC classes and their influence on the surface waterbody.

S. Mahammad, A. Islam, *Palaeochannels and Water Resource Management*,
https://doi.org/10.1007/978-3-031-92021-9_4

4.1.1 Dataset and Methods

This analysis used satellite images, which are available in the Google Earth Engine (GEE), including cloud-free satellite images of Landsat 5 of the United States Geological Survey (USGS) and Sentinel 2 of the European Space Agency (ESA). GEE is a cloud-based platform, which allows the assessment of the datasets in online mode, which means there is no need to download the raw datasets. The study used Tier 1 top of atmosphere (TOA) reflectance product of the thematic mapper (TM) of Landsat 5 and multi-spectral images (MSI) of Sentinel 2, spanning from 1989 to 2020. The study portrays the dynamics of surface features in both the pre-monsoon (PRM) and post-monsoon (POM) periods. Hence, for the PRM period, satellite data from April to June and for the period POM, satellite data from November to December months have been used (Table 4.1).

In this study, a robust methodology has been implemented, which includes the pre-processing of the datasets, supervised LULC classifications based on the three widely used machine learning (ML) models, such as random forest (RF), support vector machine (SVM), and classification and regression tree (CART), accuracy assessment of the image classification, and the intensity analysis of LULC classes. Based on the accuracy assessment of ML algorithms, the best-fit model has been taken for supervised LULC classification. The RF algorithm propped by Breiman (2001) is an efficient technique for supervised LULC classification. The RF algorithm is robust to the outlier and capable to classify images with high accuracy even in noisy environments (Binh et al., 2021). Besides, the multiple decision tree used in the RF algorithms increases the model accuracy (Chen et al., 2011). In this analysis, RF classifier with 500 trees has been implemented for the supervised classification. The CART algorithm is a predictive and nonparametric technique (Chang & Chien, 2013), which has been used in various fields. It uses the Gini index to split the root nodes into two nodes depending on the best attribute and threshold value until the last pure subset is attained in the tree (Pathak et al., 2018). Besides, it operates more accurately from the effect of the outlier of the input datasets. In this analysis, the CART algorithm has been used for LULC classification. Moreover, the SVM algorithm is useful in non-linear and larger dimensions of datasets (Ghosh

Table 4.1 Description of the satellite images used for LULC classification

Satellites and sensors	Resolution (m)	Years	Periods	Start date	End date
Landsat 5 TM (Tier 1 TOA)	30	1989	PRM	1989-02-20	1989-04-30
			POM	1999-11-01	2000-02-28
		1999	PRM	1999-02-01	1999-04-30
			POM	1999-11-01	2000-02-28
		2010	PRM	2010-02-15	2010-04-30
			POM	2010-11-01	2010-12-31
Sentinel 2 (MSI) (Tier 1 TOA)	10	2020	PRM	2020-03-01	2020-04-30
			POM	2020-11-01	2020-12-31

Source: Computed by the authors, 2022 based on Google Earth Engine datasets et al., 2019). The efficiency of the SVM is based on the maximum length from the hyperplane to the maximum margin of the classes (Warmuth et al., 2003). In this study, the SVM algorithm has been used with radial basis function (RBF) kernel due to its ability to overcome space complexity problems (Wang et al., 2016a, b).

The model validation of the LULC classification has been implemented based on the error matrix. The error matrix has been generated from the correlation between class level according to the classification and the actual class of reality (Feizizadeh et al., 2021). Based on the error matrix, user accuracy, producer accuracy, and overall accuracy can be generated. In this study, the accuracy assessment of the produced LULC maps has been executed based on overall accuracy and Kappa statistics (Eqs. 4.1 to 4.4).

$$\text{User's accuracy} = \frac{\text{Number of correct pixels in each category}}{\text{Total number of pixels in that categoty}\left(\text{the row total}\right)} \times 100 \tag{4.1}$$

$$\text{Producer's accuracy} = \frac{\text{Number of correct pixels in each category}}{\text{Total number of pixels in that categoty}\left(\text{the column total}\right)} \times 100 \tag{4.2}$$

$$\text{Overall accuracy} = \frac{\text{Number of all correct pixels}\left(\text{Diagonal total}\right)}{\text{Total number of reference pixels}} \times 100 \tag{4.3}$$

$$\text{kappa} = \frac{\left(\text{TS} \times \text{TCS}\right) - \Sigma\left(\text{Column total} \times \text{Row total}\right)}{\text{TS}^2 - \Sigma\left(\text{Column total} - \text{Row total}\right)} \times 100 \tag{4.4}$$

where TS is the total sample, and TCS is the total correctly classified sample.

The study also assessed the intensity analysis of the LULC classes using a transitional probability matrix. The intensity analysis has been done in terms of interval intensity, categorical intensity and transition intensity for both the pre-monsoon and post-monsoon season of the three intervals (1989–1999, 1999–2010, and 2010–2020). The interval intensity deals with annual changes in the areal extension of LULC classes. Therefore, it shows the rate of changes in land transformation whether it is fast or slow with respect to overall changes. Besides, the categorical intensity depicts the changes in LULC classes with their gain and loss conditions. It indicates which category is gaining its area and which category is losing its areal extension. Moreover, the transition level intensity portrays the contribution of LULC classes to the changes of the one. In this study, transition level intensity analysis has been used to find out the influence of LULC classes in terms of physical and socioeconomic activities on the dynamics of surface waterbodies. The detailed methodology of the dynamics of surface waterbody has been shown in Fig. 4.1.

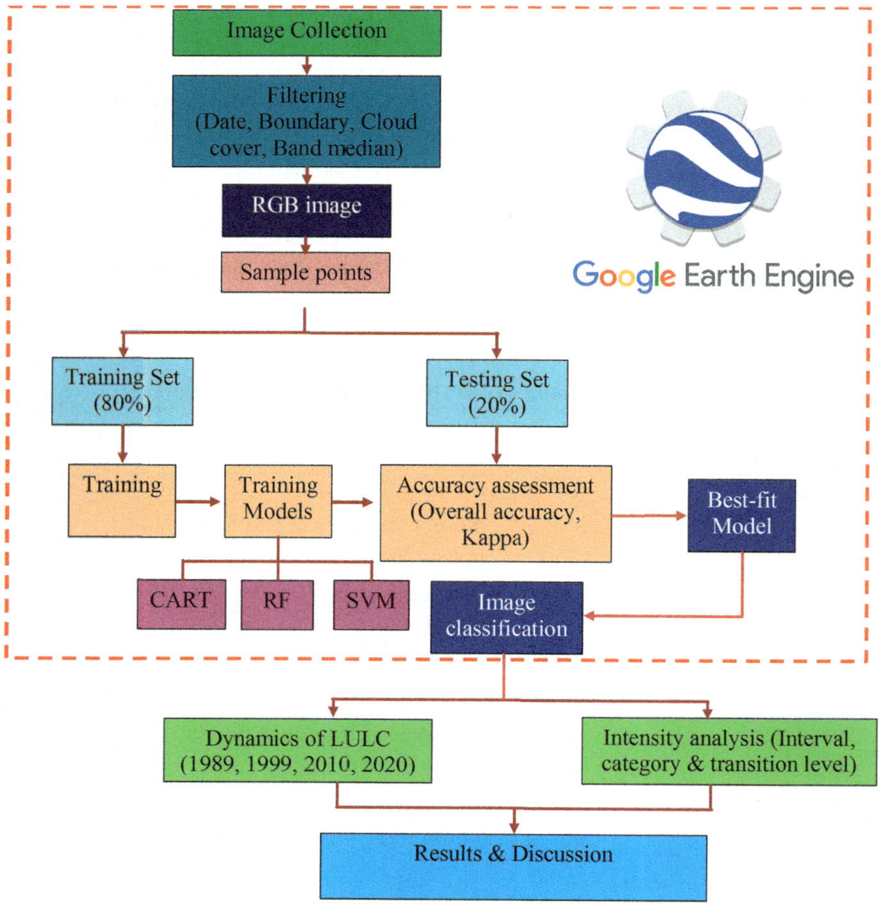

Fig. 4.1 Methodological flowchart of dynamics of the surface waterbodies

4.1.2 Results

This study focuses on the dynamics of surface waterbody and their transformations to the non-water classes. In this regard, LULC classification has been done using the best-fit ML algorithm among RF, CART, and SVM. LULC classification has been made for both the PRM and POM seasons of 1989, 1999, 2010, and 2020 along with the six major classes, including waterbodies, vegetation cover, agricultural land, agricultural fallow land, sand area, and built-up area. The result reveals that the waterbody class comprises 79.86 km^2 in the PRM period of 1989 and 85.38 km^2 in 2020. In addition, the areal extension of the surface waterbody covers 90.69 and 85.38 km^2 in 1999 and 2020, respectively (Fig. 4.2). Therefore, it is observed that the waterbody in the PRM period fluctuated by 5.53 km^2 (0.17%) from 1989 to 2020

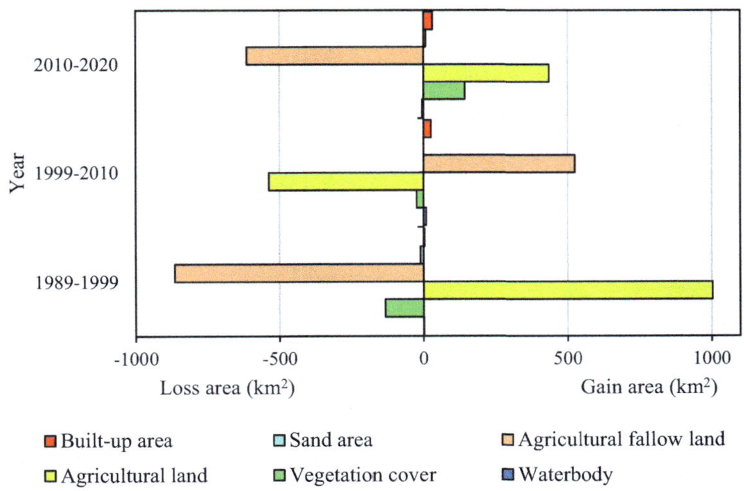

Fig. 4.2 Loss and gain area of different LULC classes in PRM season

Table 4.2 Areal extension of the LULC classes during different periods

Years	Seasons	Waterbody (km^2)	Vegetation cover (km^2)	Agricultural land (km^2)	Agricultural fallow (km^2)	Sand area (km^2)	Built-up area (km^2)
1989	PRM	79.86	601.33	499.85	1981.23	25.14	22.02
	POM	110.44	868.34	1014.78	1173.27	15.29	27.29
1999	PRM	80.97	467.56	1503.37	1118.10	13.86	25.54
	POM	104.63	496.70	1270.58	1293.99	13.98	29.53
2010	PRM	90.69	443.70	965.96	1643.10	14.30	51.65
	POM	99.31	546.83	1617.25	874.09	14.41	57.51
2020	PRM	85.38	591.02	1420.16	1000.90	19.51	92.43
	POM	96.76	685.67	1523.10	751.62	41.99	110.26

Source: Computed by the authors, 2022

(Table 4.2). Moreover, in the POM season, the spatiality of the waterbody varied significantly from 1989 to 2020. In 1989, the waterbody class covers 110.44 and 104.63 km^2 in 1999, whereas it comprises 99.31 and 96.76 km^2 in 2010 and 2020, respectively (Fig. 4.3). Therefore, the declining trend of waterbody has been found in −13.68 km^2 (−0.43%) from 1989 to 2020.

The vegetation cover of the study area included natural vegetation, plantation, shrubs, and orchards. The vegetation cover of an area plays a significant role to protect soil erosion and maintain environmental balance. As per LULC classification, the loss of vegetation has been found in both the PRM and POM periods. In the PRM season, vegetation cover accounted for 601.33 km^2 in 1989, which decreases to 467.56 in 1999 and 433.7 km^2 in 2010. However, it increases to 591.02 km^2 in 2020. Therefore, 10.31 km^2 (0.32%) loss of vegetation cover has been observed in the PRM season from 1989 to 2020. Similar to the PRM season, the vegetation

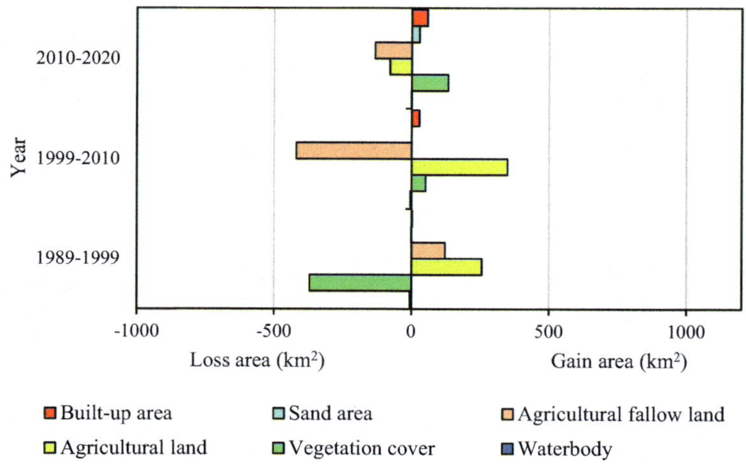

Fig. 4.3 Loss and gain area of different LULC classes in the POM season

cover in the POM season recorded decreasing trend of −5.69% from 1989 to 2020 (Table 4.2).

In this study, the LULC of class agricultural land entails the cultivated area or the agricultural land having crops during the classification. The results depict that the area of agricultural land in both the periods of PRM (Fig. 4.4) and POM (Fig. 4.5) recorded significant growth from 1989 to 2020. In the PRM period, the cropland area covered 499.85 km^2, which increased to 965.96 km^2 in 2010. After that, the cropland occupied the areal extension of 1420.16 km^2 in 2020 with 28.7% increase in area from 1989. Like the PRM season, the spatial distribution of the agricultural land shows a significant growth of 31.62 km^2 (15.84%) in the POM season from 1989 to 2020. Apart from that, agricultural land extended over 1014.78, 1270.58, 1617.25, and 1523.1 km^2 in 1989, 1999, 2010, and 2020, respectively.

The agricultural fallow is associated with agricultural lands, which are temporarily uncultivated for one or more seasons. The agricultural fallow land, after one or more seasons, remains temporarily uncultivated depending on the soil characteristic, availability of water resources, and demand for the crops in the study area. Based on the LULC classification, the estimated extent of agricultural fallow land for the PRM period is as follows: 1981.23, 1118.1, 1643.1, and 1000.9 km^2 in 1989, 1999, 2010, and 2020, respectively. Besides, in the POM period, a decreasing trend of agricultural fallow land has been observed. The areal extent of agricultural fallow occupied 1173.27, 1293.99, 874.09, and 751.62 km^2 in the POM period of 1989, 1999, 2010, and 2020, respectively.

The sand area includes a sand bar along the river channel, sandy exposures, and bare soil, which generally do not have vegetation cover in the study area. Based on the LULC classification, the areal extent of the sand area was observed as 25.14, 13.86, 14.30, and 19.51 km^2 in the PRM period of 1989, 1999, 2010, and 2020.

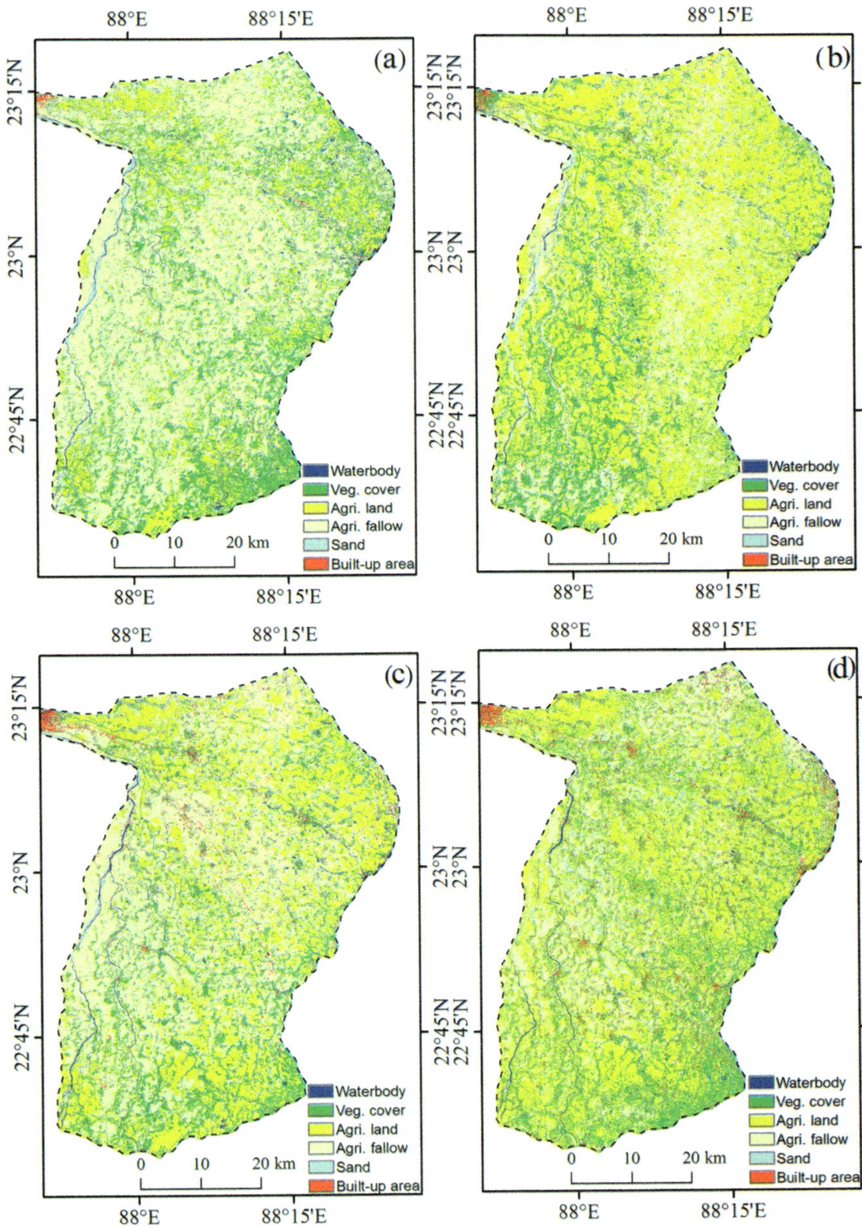

Fig. 4.4 Spatial distribution of LULC in the PRM period: (**a**) 1989, (**b**) 1999, (**c**) 2010, and (**d**) 2020

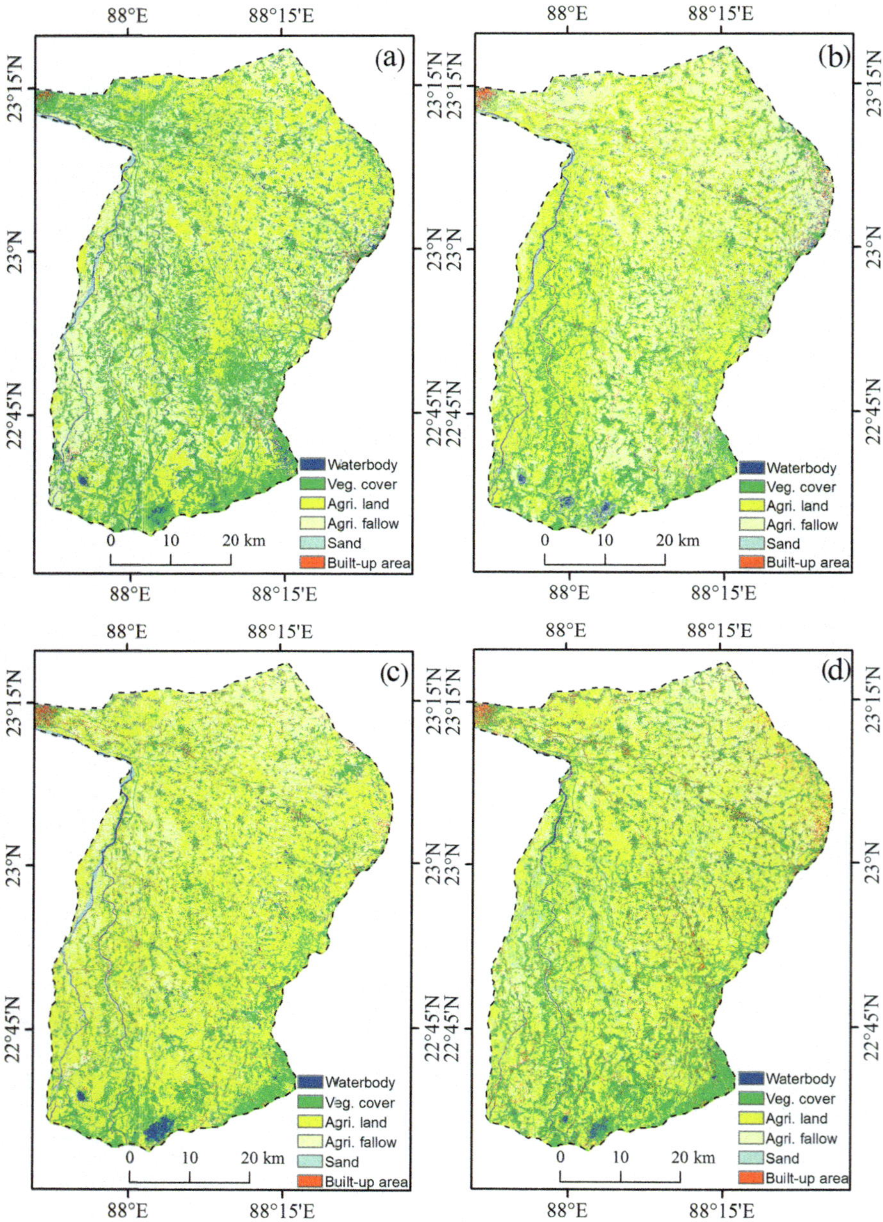

Fig. 4.5 Spatial distribution of LULC in POM period: (**a**) 1989, (**b**) 1999, (**c**) 2010, and (**d**) 2020

Therefore, the decreasing trend of the sand area has been observed in the PRM season. Besides, the sand area in the POM season is estimated to be covered 15.29, 13.98, 14.41, and 41.99 km^2 in 1989, 1999, 2010, and 2020. Moreover, the increasing trend of the sand area (0.83%) has been observed from 1989 to 2020.

The built-up area of this study included settlement areas, roads, railways, industrial areas, and constructions related to livelihood. The LULC classification reveals that the built-up area extent was 22.02, 25.54, 51.65, and 92.43 km^2 in 1989, 1999, 2010, and 2020, respectively, for the PRM period. The areal extent of built-up area occupied 27.29, 29.53, 57.51, and 110.26 km^2 in the POM period of 1989, 1999, 2010, and 2020, respectively. Therefore, the increasing trend of the built-up area has been observed in both the PRM and POM periods.

4.1.2.1 Intensity Analysis

In this research, intensity analysis has been conducted in terms of interval intensity, categorical intensity, and transition intensity to detect the dynamics of LULC.

Interval Level Intensity

The changing intensity is higher in both periods of PRM and POM from the assumed uniform intensity. In the PRM season, the rate of changes in 1989–1999 is 27.41 km^2 per year indicating a fast rate of change (Fig. 4.6a). The change rate of 1999–2010 is the lowest in terms of annual changes (21.54 km^2/year) revealing the slow rate of land transformation. Besides, the annual change rate of 2010–2020 is observed near the same rate of uniform intensity. Apart from that, a similar trend of annual changes has been found in the POM season (Fig. 4.6b). In this period, the highest rate of annual changes has been found in first-time intervals (1989–1999) with an annual change rate of 32.66 km^2/year, which infers the fast change of LULC classes. Moreover, the time intervals of 1999–2010 and 2010–2020 account for the annual change rate of 23.19 and 24.40 km^2/year, respectively, which depicts the slow changes from the uniform intensity. The trend has also been found in the study of LULC dynamics by Manzoor et al. (2022).

Category Level Intensity

The categorical intensity has been analysed accompanied by gain intensity, loss intensity, and uniform intensity. Uniform intensity reveals the average growth rate of LULC categories (Manzoor et al., 2022). It reveals that if a gain class crosses the uniform intensity, it is considered an active gainer and when the gain class is located below the uniform intensity, it is known as a dormant gainer. Similarly, the loss class crossing the uniform intensity is categorized as an active loser and when located below the uniform intensity, it is designated as a dormant loser class. In this study

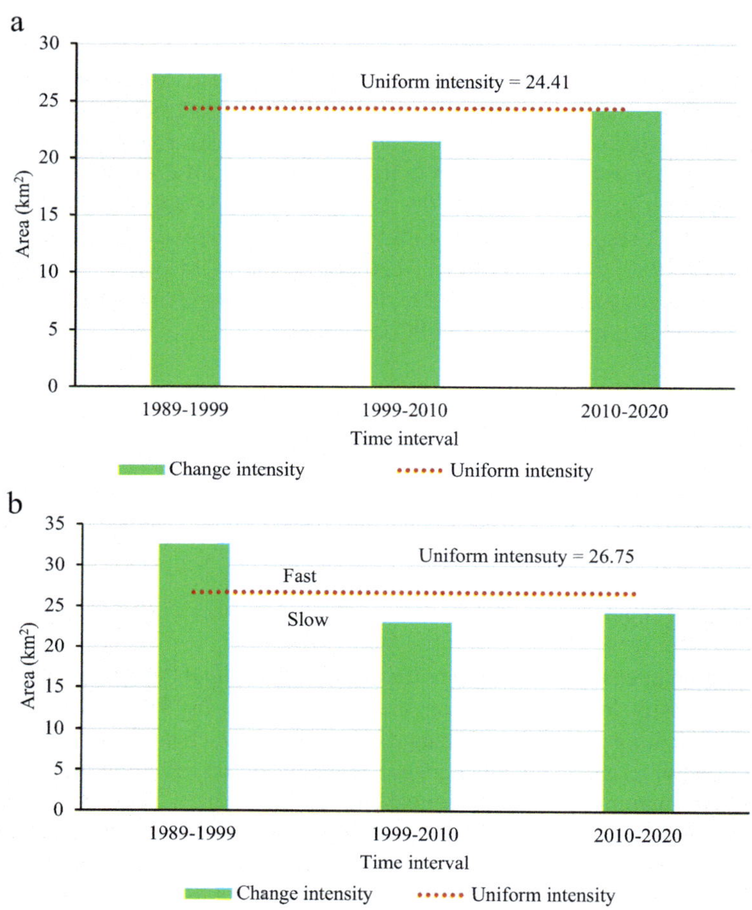

Fig. 4.6 Interval intensity analysis: (**a**) PRM and (**b**) POM

area, vegetation cover, agricultural land, and agricultural fallow represent the active classes in terms of loss and gain category for both the PRM and POM periods during the study period. In the PRM, except the last time interval, the vegetation cover is the active loser and contributes to the agricultural land and agricultural fallow land during 1989–1999 and 1999–2010 time intervals (Fig. 4.7a–c). In the POM period, vegetation covers are the active loser in 1989–1999; however, it is the active gainer in the time interval of 1999–2010. In the time interval of 2010–2020 in PRM and POM seasons, vegetation cover has been considered as active gainer though its loss intensity is closer to the uniform intensity during this time interval. Therefore, the results suggest that vegetation cover in the study area is gradually decreasing and contributes to the agricultural land, agricultural fallow, and built-up area. Moreover, agricultural land and agricultural fallow land demarcated active loss and gain category various times during the study period in both PRM and POM seasons.

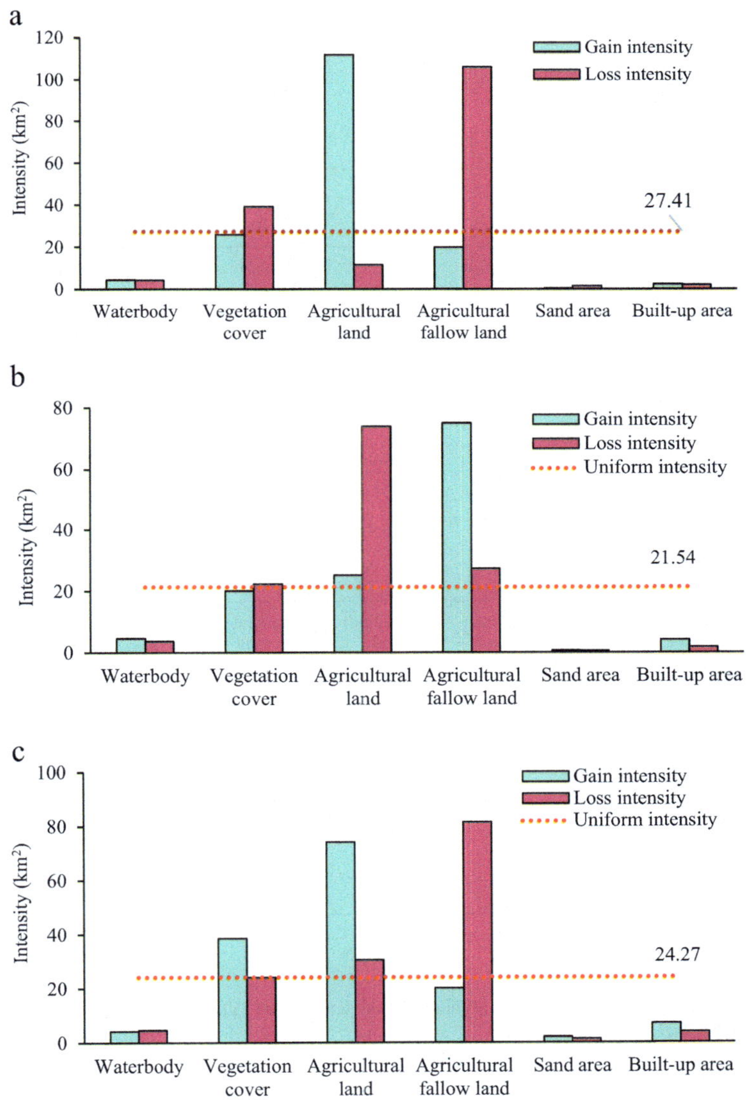

Fig. 4.7 Category level intensity analysis of PRM season: (**a**) 1989–1999, (**b**) 1999–2010, and (**c**) 2010–2020

Therefore, these land use categories dominantly alter the LULC classes in the study area. The agricultural land and agricultural fallow land are alternately used in the PRM and POM periods for cultivation purposes. Additionally, it is observed that waterbody, built-up area, and sand area are considered dormant classes in terms of gain and loss intensity (Fig. 4.8a–c). However, the increasing rate of loss and gain intensity of these classes has been observed during the study intervals.

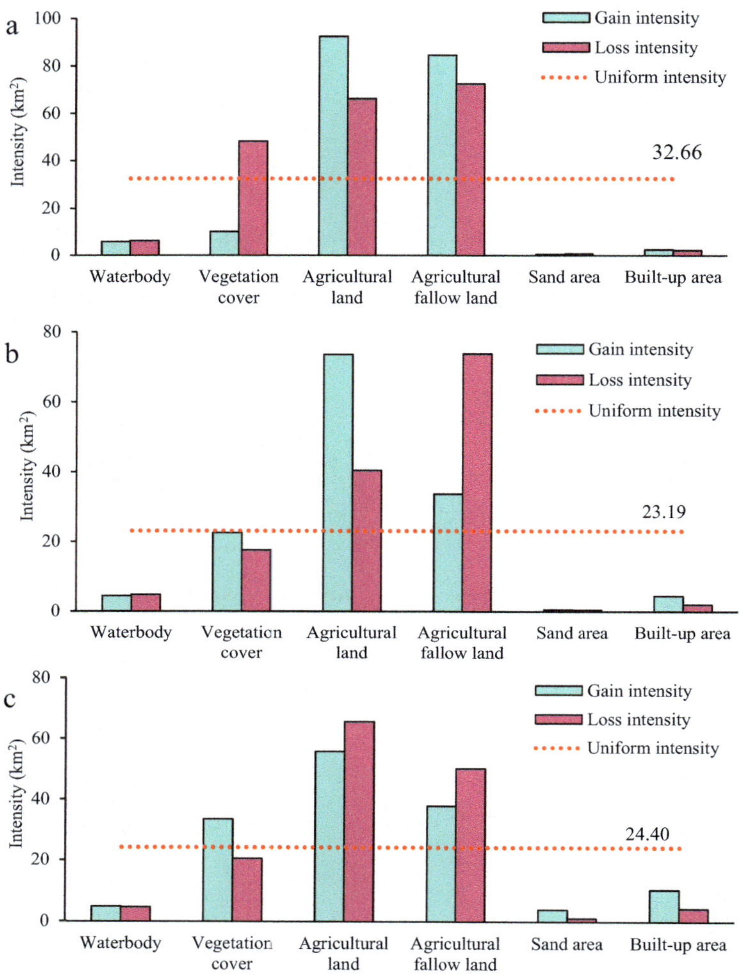

Fig. 4.8 Category level intensity analysis of POM season: (**a**) 1989–1999, (**b**) 1999–2010, and (**c**) 2010–2020

Transition Level Intensity

In this research, transition level intensity for waterbody has been analysed for both PRM and POM seasons across three intervals, i.e. 1989–1999, 1999–2010, and 2010–2020. It is observed that the dynamics of waterbody is significantly controlled by the vegetation cover, agricultural land, and agricultural fallow land for both the PRM and POM seasons. In the PRM period of 1989–1999, it was found that the agricultural fallow land (2.53 km^2) and vegetation cover (1.13 km^2) have a larger transition intensity value than the uniform intensity (0.88 km^2) (Fig. 4.9a). Therefore, it indicates the transformation of the waterbody is largely affected by agricultural activities and vegetation covers. In the second interval (1999–2010), it is found that

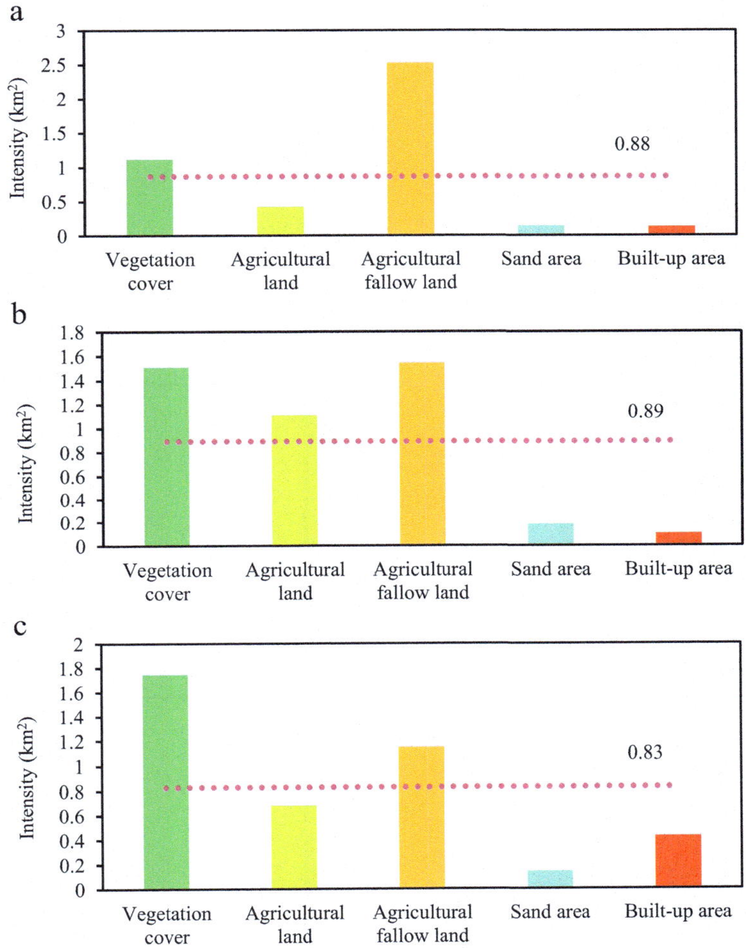

Fig. 4.9 Transition intensity for waterbody of PRM season: (**a**) 1989–1999, (**b**) 1999–2010, and (**c**) 2010–2020

the agricultural fallow land (1.54 km^2), agricultural land (1.12 km^2), and vegetation cover (1.51 km^2) have the higher contribution to the dynamics of the waterbody (6.9b). Apart from that, in the third interval (2010–2020), agricultural fallow land (1.15 km^2) and vegetation cover (1.75 km^2) have larger transition intensity values than the uniform intensity (0.83 km^2) (Fig. 4.9c). Besides, in the POM season of the first interval, vegetation cover has been considered the dominant active contributor to the waterbody having an intensity value of 3.07 km^2, which is higher than the uniform intensity (Fig. 4.10a). In the second interval, agricultural fallow and agricultural land are considered the dominant active contributors to changes in the waterbodies (Fig. 4.10b). Besides, in the third interval, vegetation cover is the

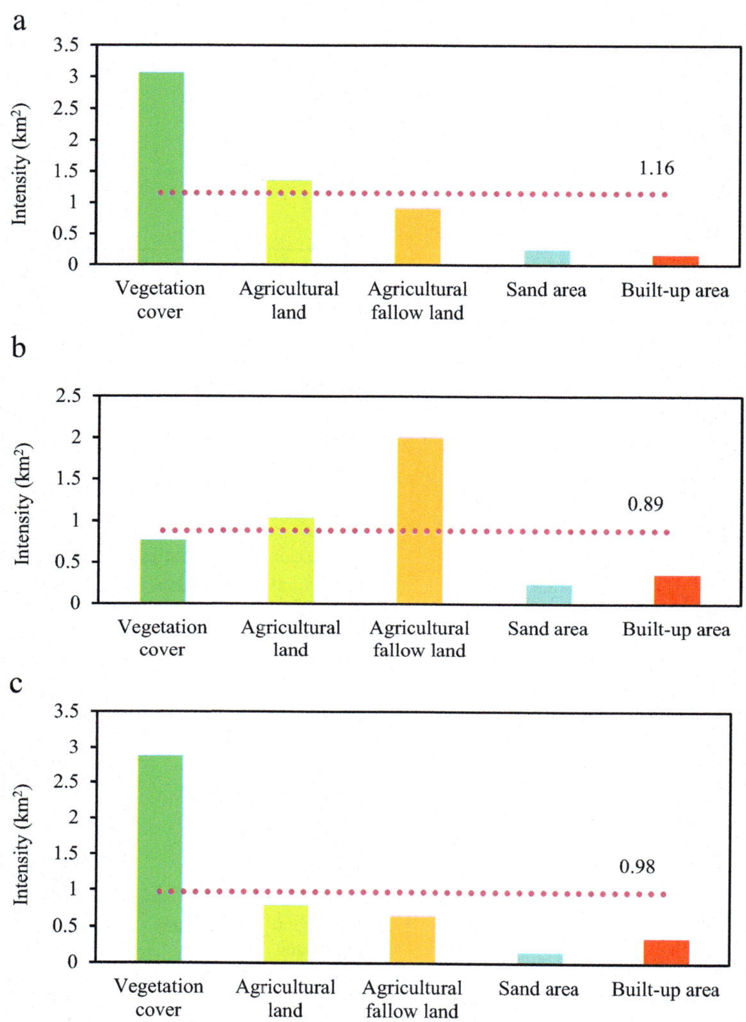

Fig. 4.10 Transition intensity for waterbody of POM season: (**a**) 1989–1999, (**b**) 1999–2010, and (**c**) 2010–2020

highest contributor to the dynamics of the waterbody in the study area. The result illustrated that in terms of agricultural fallow land, vegetation cover and agricultural fallow land contribute significantly to the dynamics of the waterbody (Fig. 4.10c). It implies that seasonal agricultural activity occurs in the waterbody area. The result depicts the season vegetation cover growth in the waterbody class of the study area.

4.1.2.2 Accuracy Assessment and Critical Analysis

The three MK classifiers (CART, RF, and SVM) have been implemented for the classification of the satellite images. A confessional matrix has been developed to calculate the user accuracy (UA) and the producer accuracy (PA) based on the training and testing sample points. After that, overall accuracy (OA) and Kappa coefficient have been defined. The best-fit algorithm has been used for image classification depending on the highest value of OA and Kappa of the period. The accuracy assessment of the ML algorithms reveals that the RF classifier scored highest in terms of OA and Kappa coefficient of all the seasons indicating the best-fit classifier for image classification (Table 4.3). Therefore, the supervised image classifications of all seasons from 1989 to 2020 have been done using an RF classifier.

This study reveals that the alteration of LULC is not consistent in the three interval periods, indicating changing characteristics of the factors responsible for the alteration of LULC in the study area. The physical and socio-economic factors, including topographic condition, climatic characteristics, population growth, as well as economic growth are considered to be the dominant drivers of the transformation of the LULC types (Abbas et al., 2022). However, the impact of socio-economic factors may reshape the LULC types more rapidly than the physical factors. The supervised LULC classification reveals the reduction of the waterbody, vegetation cover, and agricultural fallow land and the increase in agricultural land and built-up area during the study periods. In the PRM period, the expansion of the waterbody class fluctuates from 1989 to 2020. Moreover, in the POM period, the area covered by waterbody is gradually decreasing from 1989 to 2020. The seasonal distribution of rainfall is closely associated with the distribution of the waterbodies in the study area (Fig. 4.11). The rainfall distribution of the PRM period (April–June) has a positive trend ($r = 0.69$). Additionally, the rainfall distribution of the POM *kharif* season (January–March) has a positive trend ($r = 0.86$). Therefore, the increasing trend of waterbody of the PRM season is closely associated with the rainfall of the POM and POM *kharif* seasons in the study area. Apart from that, the

Table 4.3 Accuracy assessment of the ML classifier in terms of overall accuracy and kappa statistics

		Overall accuracy			Kappa statistics		
Year	Season	CART	RF	SVM	CART	RF	SVM
1989	PRM	0.811	0.838	0.750	0.748	0.778	0.639
	POM	0.908	0.939	0.847	0.902	0.910	0.810
1999	PRM	0.837	0.908	0.765	0.791	0.882	0.693
	POM	0.879	0.885	0.739	0.834	0.840	0.602
2010	PRM	0.845	0.904	0.769	0.799	0.876	0.685
	POM	0.840	0.885	0.669	0.817	0.877	0.544
2020	PRM	0.773	0.777	0.770	0.715	0.758	0.693
	POM	0.735	0.788	0.657	0.623	0.751	0.607

Source: Computed by the authors, 2022

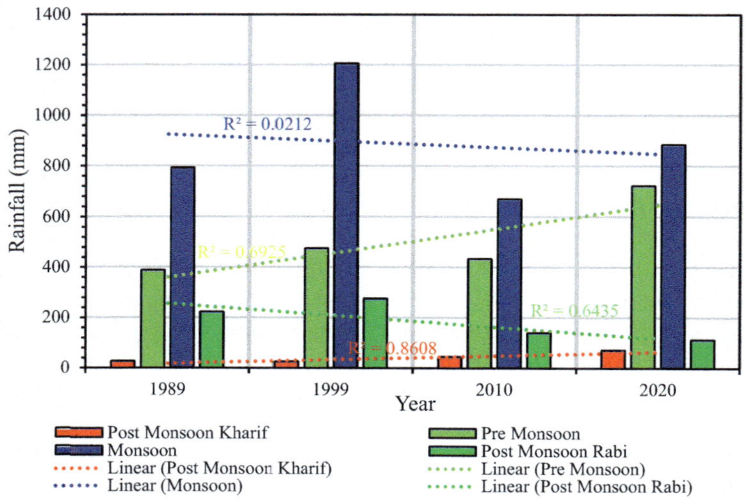

Fig. 4.11 Distribution of seasonal rainfall of 1989, 1999, 2010, and 2020 (based on CHIRPS datasets)

spatial extension of waterbodies is reduced from 1989 to 2020. The rainfall distribution of monsoon (MON) (July–September) and POM *rabi* season (October–December) indicates a declining trend suggesting the reduction of waterbodies. Moreover, the eutrophication and sedimentation of stagnant waterbodies and the filling up of ponds are responsible for transforming waterbodies in the study area.

Like the waterbody, the vegetation cover of the study area has been experiencing decreasing trend from 1989 to 2020. The declination of vegetation cover in the study area has been reported by Dutta et al. (2020). The reduction of vegetation cover has occurred through the expansion of the built-up areas and agricultural land (Rahaman et al., 2020). The results of the study reveal that the declination of vegetation cover is found in higher magnitude in 1989–1999. After that, in the transition of 1999–2010 and 2010–2020, covering both the PRM and POM periods, the declination of vegetation is often minimum, and its increasing trend has been observed. The reasons behind this growth of vegetation cover are the transformation of agricultural fallow and cropland into orchard cultivation (Das & Sarkar, 2019). The agricultural fallow land of the study area is rapidly decreasing due to its transformation into agricultural land, built-up areas, as well as plantation cultivation. Additionally, agricultural land and built-up area have been increased during the study period due to increasing population and socio-economic growth. The study area located in the fertile zone of West Bengal is famous for agricultural productivity. It makes the area populous with a higher growth rate which is a significant indicator for transforming the LULC classes.

4.2 Dynamics of Groundwater Level and its Future Prediction Using Machine Learning Algorithms

Because of its high quality and widespread distribution, groundwater is an essential supply of freshwater utilized for industrial, agricultural, and drinking reasons (Mahammad & Islam, 2021). It also encourages ecological sustainability, environmental management, and economic development (Kombo et al., 2020; Seifi et al. 2020). Because external forces have less of an impact on groundwater supplies than on surface water, they have increased dramatically. More groundwater is extracted in India than in any other nation in the world, with reports indicating that 89% of it is used for irrigation, 9% for residential use, and 2% for industrial uses (Margat & Van der Gun, 2013; Ahada & Suthar, 2018; Mahammad & Islam, 2021). Furthermore, the quality of groundwater is a growing problem in India due to the massive anthropogenic extraction of groundwater and the various human interventions that pollute it, resulting in increased levels of heavy metal and pollution in the groundwater. The primary problem in the context of DFD is the availability of groundwater with minimum quality. As a result, the goal of this chapter is to investigate groundwater dynamics, or the groundwater fall rate throughout the year. Additionally, this chapter provides a critical analysis of the spatial–temporal distribution of groundwater depth and their future scenario. Additionally, the quality of groundwater for irrigation and drinking has been discussed in the following sections of this chapter.

4.2.1 Datasets and Methods

The Central Ground Water Board (CGWB), a division of the Indian government, provided the groundwater level statistics for the years 2013–2014 through 2020–2021. The GWL of the available 30 wells throughout the PRM (April–June), MON (July–September), POM *rabi* (October–December), and POM *kharif* (January–March) periods has been examined in accordance with the CGWB (2021) standard (Fig. 4.12). In addition, an analysis was conducted on the elements that influence GWL changes, including population increase, rainfall, irrigation using groundwater, and the domestic and industrial sectors. Therefore, from 2013 to 2021, rainfall data was gathered using the Climate Hazards Group Infrared Precipitation with Station data (CHIRPS). Additionally, CD block-wise census data for 1961, 1971, 1981, 1991, 2001, and 2011 were gathered from the district census handbooks in order to evaluate the population growth of the study area. Aside from that, data were gathered from CGWB publications on the dynamics of groundwater resources (2004, 2009, 2011, and 2013) regarding the stage of groundwater development (SGWD) and uses of groundwater for irrigation, household, and industrial reasons. A strong technique has been used to carry out this investigation. Figure 4.13 displays a thorough flowchart of the research procedures performed in this study.

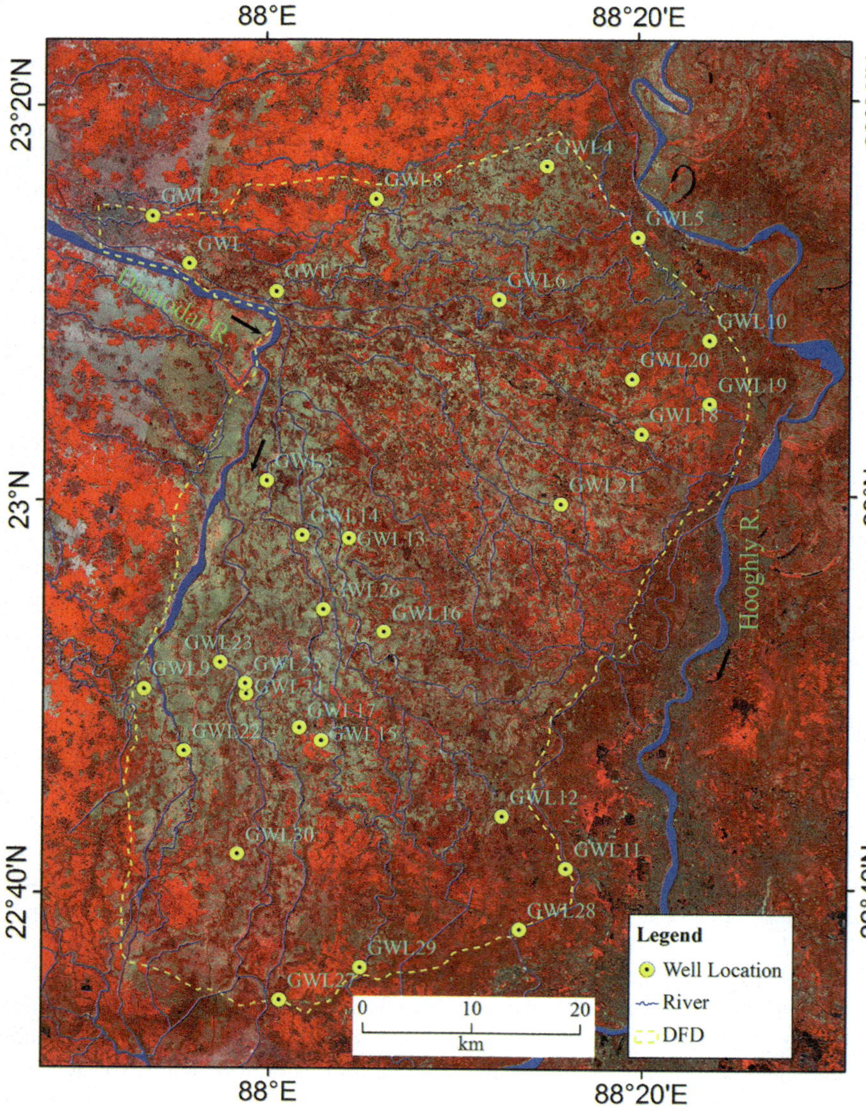

Fig. 4.12 Location of the groundwater level sample wells

4.2.2 Results

This study examines the dynamics of groundwater levels in 30 wells during the PRM, MON, POM *rabi*, and POM *kharif* seasons, spanning 2013–2014 to 2020–2021 (Table 4.4). According to the analysis, GWL varies significantly throughout the course of the years and the seasons (Fig. 4.14a). GWL is found to differ significantly in PRM, MON, POM *rabi*, and POM *kharif* seasons during the

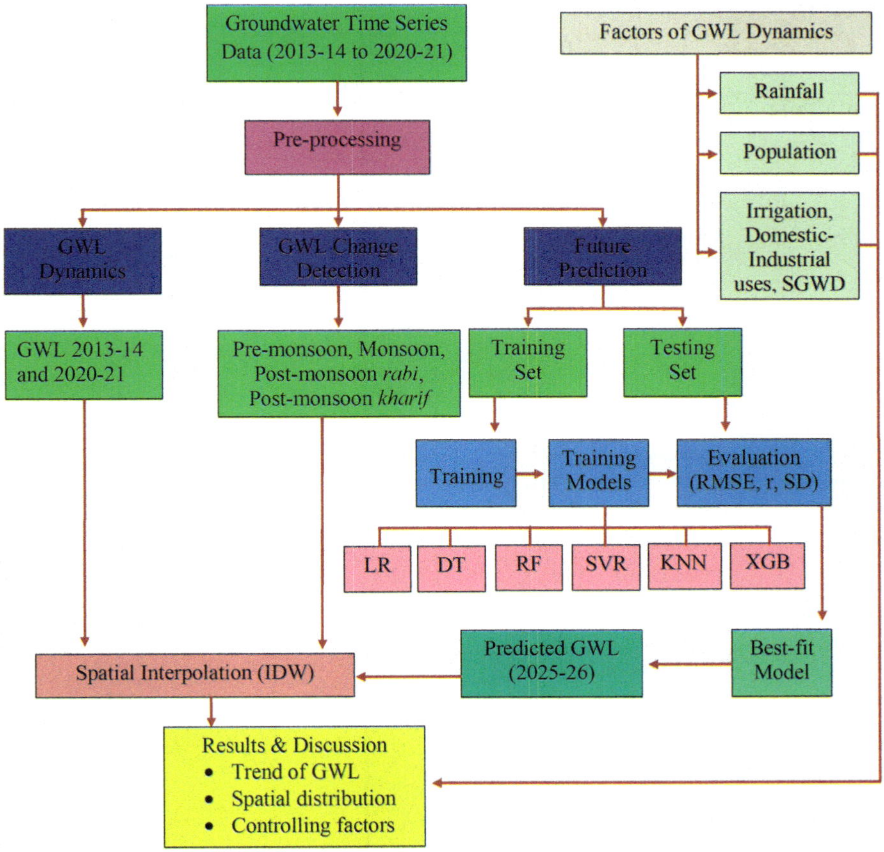

Fig. 4.13 Methodological flowchart for the detection of GWL dynamics (Based on Mahammad et al., 2023)

evaluation period (2013–2014 to 2020–2021). During the 2013–2014 PRM season, groundwater reached a maximum depth of 21.45 mbgl and a minimum depth of 7.52 mbgl, with an average of 8.25 ± 0.73 mbgl. The maximum, minimum, and average GWL values for 2020–21 were 22.35 mbgl, 7.68 mbgl, and 9.57 ± 1.89 mbgl, respectively. With the exception of the years 2014–2015, 2018–2019, and 2020–2021, the study shows that the depth of GWL increased steadily at an average rate of 0.17 m/year. With an r-value of 0.71, the trend line (Fig. 4.14b) also shows the rising rate of GWL depth. The average GWL levels varied over the MON season, ranging from 7.72 ± 0.55 mbgl to 5.67 ± 0.18 mbgl. The observed rate of change in GWL was 0.26 m/year. The trend line illustrates the GWL's growing tendency ($r = 0.81$) between 2013–2014 and 2020–2021. Furthermore, the groundwater level (GWL) dynamics throughout the POM *rabi* season (except from 2019–2020 and 2020–2021) indicate a growing rate of GWL. In 2013–2014, the average groundwater depth was 4.06 ± 1.01 mbgl, while in 2020–2021, it was

Table 4.4 Statistical summary of the groundwater levels

Year	Season	Groundwater level (mbgl)			
		Maximum	Minimum	Average	SD
2013–2014	PRM	21.45	7.52	8.25	0.73
	MON	18.43	5.49	5.67	0.18
	POM *rabi*	16.65	3.05	4.06	1.01
	POM *kharif*	19.9	5.51	6.01	0.5
2014–2015	PRM	20.6	5.9	7.47	1.57
	MON	18.14	4.25	5.96	1.71
	POM *rabi*	17.2	3.45	4.78	1.33
	POM *kharif*	17.82	3.92	7.09	3.17
2015–2016	PRM	21.76	5.05	8.14	3.09
	MON	21.76	5.14	5.9	0.76
	POM *rabi*	17.98	5.45	5.77	0.32
	POM *kharif*	18.6	5.45	7.59	2.14
2016–2017	PRM	22.04	6.25	9.84	3.59
	MON	17.89	4.97	8.27	3.3
	POM *rabi*	18.75	4.63	5.47	0.84
	POM *kharif*	19.28	7.02	9.29	2.27
2017–2018	PRM	21.2	7.45	10.14	2.69
	MON	19.54	4.79	6.09	1.3
	POM *rabi*	18.75	3.8	4.93	1.13
	POM *kharif*	18.99	6.77	7.29	0.52
2018–2019	PRM	20.44	10.6	13.37	2.78
	MON	19.87	6.58	6.93	0.35
	POM *rabi*	19.27	5.78	7	1.22
	POM *kharif*	21.42	8.6	9.81	1.21
2019–2020	PRM	22.15	13.74	13.77	0.03
	MON	19.25	9.36	10.21	0.85
	POM *rabi*	17.62	6.06	7.45	1.39
	POM *kharif*	18.82	7.25	8.71	1.46
2020–2021	PRM	22.35	7.68	9.57	1.89
	MON	19.25	7.17	7.72	0.55
	POM *rabi*	18.88	3.9	4.41	0.51
	POM *kharif*	19.6	6.73	9.03	2.3

Source: Mahammad et al., 2023

4.41 ± 0.51 mbgl. The GWL dynamics, with an r-value of 0.74, show a rising trend. During the POM *kharif* season, the average GWL increased at a maximum rate of 0.38 m/year, from 6.01 ± 0.5 mbgl in 2013–2014 to 9.03 ± 2.3 mbgl in 2020–2021. This season saw a steady increase in the GWL rate, with the exception of 2017–2018. The trend line shows that from 2013–2014 to 2020–2021, there was an increasing tendency in the depth of GWL ($r = 0.78$).

Thirty wells' groundwater levels from the PRM, MON, POM *rabi*, and POM *kharif* seasons in 2013–2014 and 2020–2021 were analysed in terms of their spatial distribution. Furthermore, measurements were made of the seasonal variation

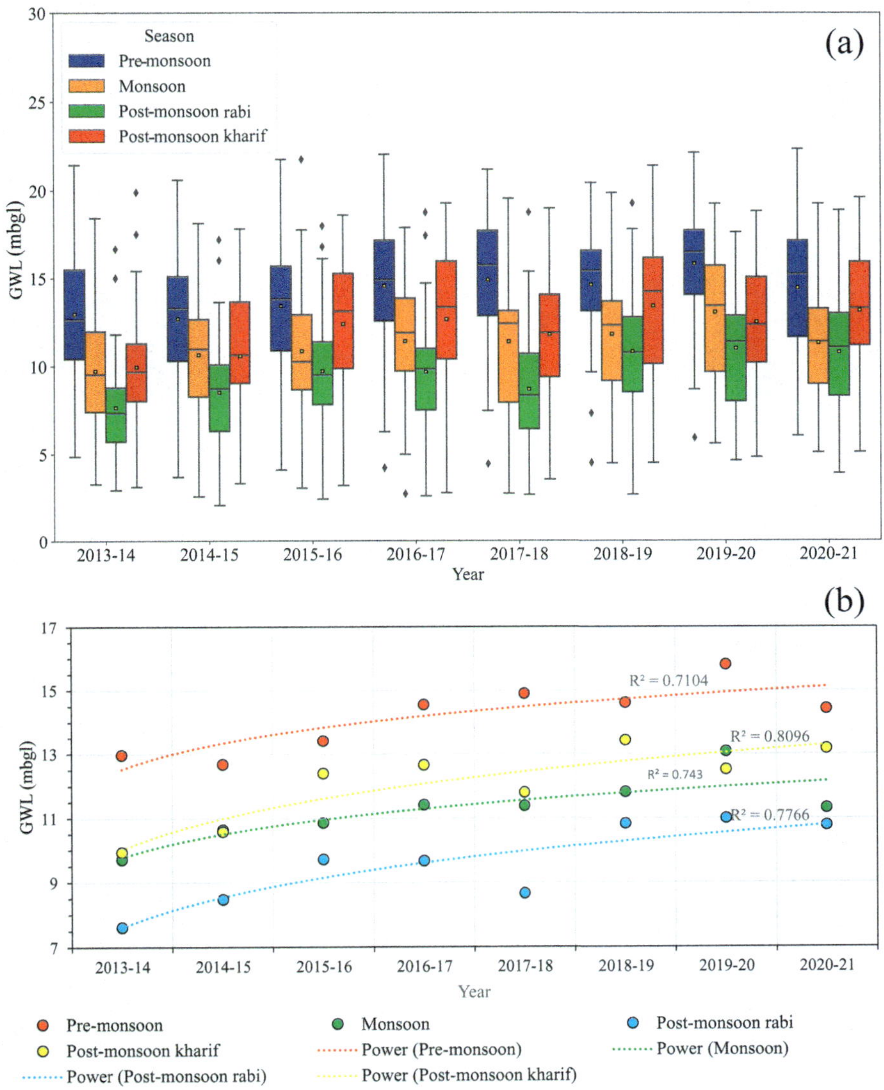

Fig. 4.14 Dynamics of groundwater level: (**a**) Distribution of groundwater levels in different periods from 2013–2014 to 2020–2021 and (**b**) Temporal change in the GWL in PRM, MON, and POM *rabi* and *kharif* periods (Based on Mahammad et al., 2023)

between the PRM and POM *rabi* seasons. The outcome shows that during the PRM season of 2013–2014, the mean GWL was 8.25 ± 0.73 (mbgl), with a maximum of 21.45 (mbgl) and a minimum of 7.52 (mbgl) (Table 4.4). The GWL's spatial distribution shows that the research area's northern and southern borders, as well as the western region around Barddhaman, are where the lowest GWL was found. The deepest GWL was found around Memari and Pandua CD blocks (Fig. 4.15).

During this time, 1.77 km^2 (0.06%) of the areal extension was occupied by a single well with a GWL <5 mbgl. A region of 249.28 km^2 (7.77%), located in the

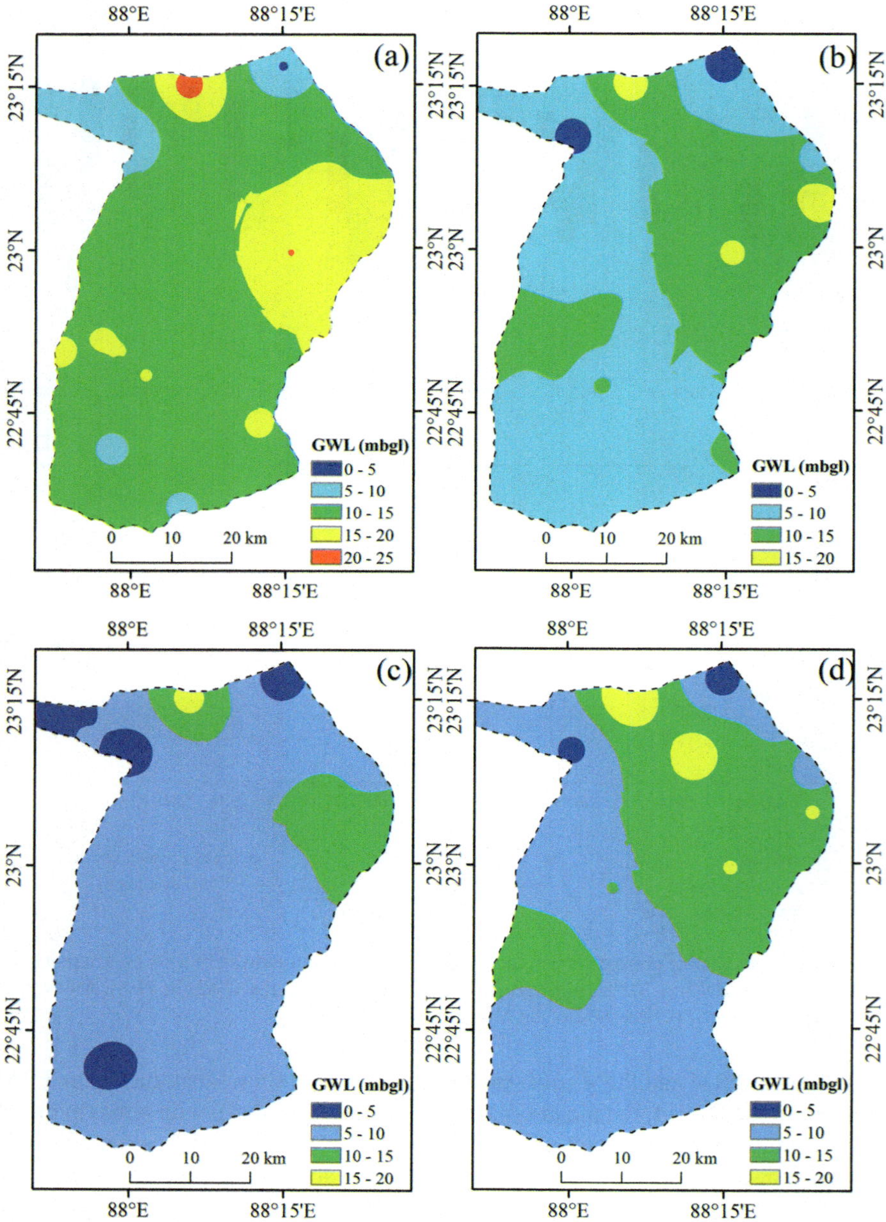

Fig. 4.15 Dynamics of groundwater level in 2013–2014: (**a**) PRM, (**b**) MON, (**c**) POM *rabi*, and (**d**) POM *kharif* season (Based on Mahammad et al., 2023)

research area's western, northern, and southern patches, was included in the GWL of 5–10 mbgl. Furthermore, it was discovered that 2296 km^2 (71.57%) of the research region, with the exception of the eastern, western, northern, and southern peripheral areas, lie within the GWL of 10–15 mbgl. Additionally, it was discovered that the GWL between 15 and 20 mbgl covered 646.54 km^2 (20.15%) in the east of the research region. The northern portion, covering 14.34 km^2 (0.45%), was home to the deepest GWL (≥20 mbgl). The GWL was significantly lower during the MON season, ranging from 5.49 to 18.43 mbgl with a mean of 5.67 ± 0.18 mbgl. This season, GWL class <5 mbgl was seen in Simlon and Pallaroad village, covering the area of 52.74 km^2 (1.64%) of the research region, while GWL of 15–20 mbgl was detected around Parahati, Pandua, and Mohanad villages, covering 68.49 km^2 (2.14%) (Table 4.4). Furthermore, class 10–15 mbgl spanned an area of 1215.74 km^2 (37.9%), whereas class 5–10 mbgl occupied the highest areal extension, accounting for 1870.96 km^2 (58.32%). During the 2013–2014 POM *rabi* season, the GWL averaged 4.06 ± 1.01 mbgl, with a range of 3.05 to 16.65 mbgl. This was the season with the lowest GWL out of the four. The research area was most fully covered in this season by the GWL class 5–10 mbgl, which expanded across 2667.97 km^2 (83.17%), followed by class 10–15 mbgl, which covered 342.67 km^2 (10.68%), and the remaining classes, which were restricted to a small region. In comparison to the POM *rabi* season, the GWL increased in the POM *kharif* season. GWL varied from 5.51 to 19.9 mbgl throughout this season, averaging 6.01 ± 0.5 mbgl. The GWL's spatial pattern had four classes: the <5 mbgl class covered the least amount of 39.63 km^2 (1.24%) in the research region, while the 5–10 mbgl class accounted for the highest area, or 1769.54 km^2 (55.16%). In addition, 1290.38 km^2 (40.22%) of the study area was inhabited by the 10–15 mbgl class and 108.39 km^2 (3.38%) by the 15–20 mbgl class. During the MON, POM *rabi*, and POM *kharif* seasons, there was no presence of the GWL class of ≥20 mbgl.

For 2020–2021, the mean groundwater depth during the PRM season was 9.57 ± 1.89 mbgl, with a high groundwater level of 22.35 mbgl and a lowest groundwater depth of 7.68 mbgl (Table 4.4). The deepest groundwater level was found to have grown over the Bara Dhamas and Mahanad area, whereas the western portion of the study area showed the lowest groundwater depth, according to the spatial distribution of the GWL (Fig. 4.16). Not a single well during this time had a GWL less than 5 mbgl. The GWL of 5–10 mbgl was found in the western regions of the research area, covering an area of 137.18 km^2 (4.28%). Furthermore, the GWL of 10–15 mbgl including the northeastern, western, and southern regions of the study area is explained by an area of 1410.28 km^2 (43.96%). Furthermore, it was discovered that the central and eastern portion of the study region was covered by the GWL of 15–20 mbgl, which covered 1586.83 km^2 (49.47%). A total of 73.64 km^2 (2.3%) was included in the GWL (≥20 mbgl) (Table 4.5). With a mean of 7.72 ± 0.55 mbgl, the GWL varied from 7.17 to 19.25 mbgl over the MON season. In this season, it was discovered that the GWL of 5–10 mbgl covered an area of 672.92 km^2 (20.98%), which was distributed over the northern, southern, and western regions. The central region of the research area, comprising 2351.46 km^2 (73.3%), was covered by the GWL of the 10–15 mbgl class. Furthermore, a GWL of 15–20 mbgl was

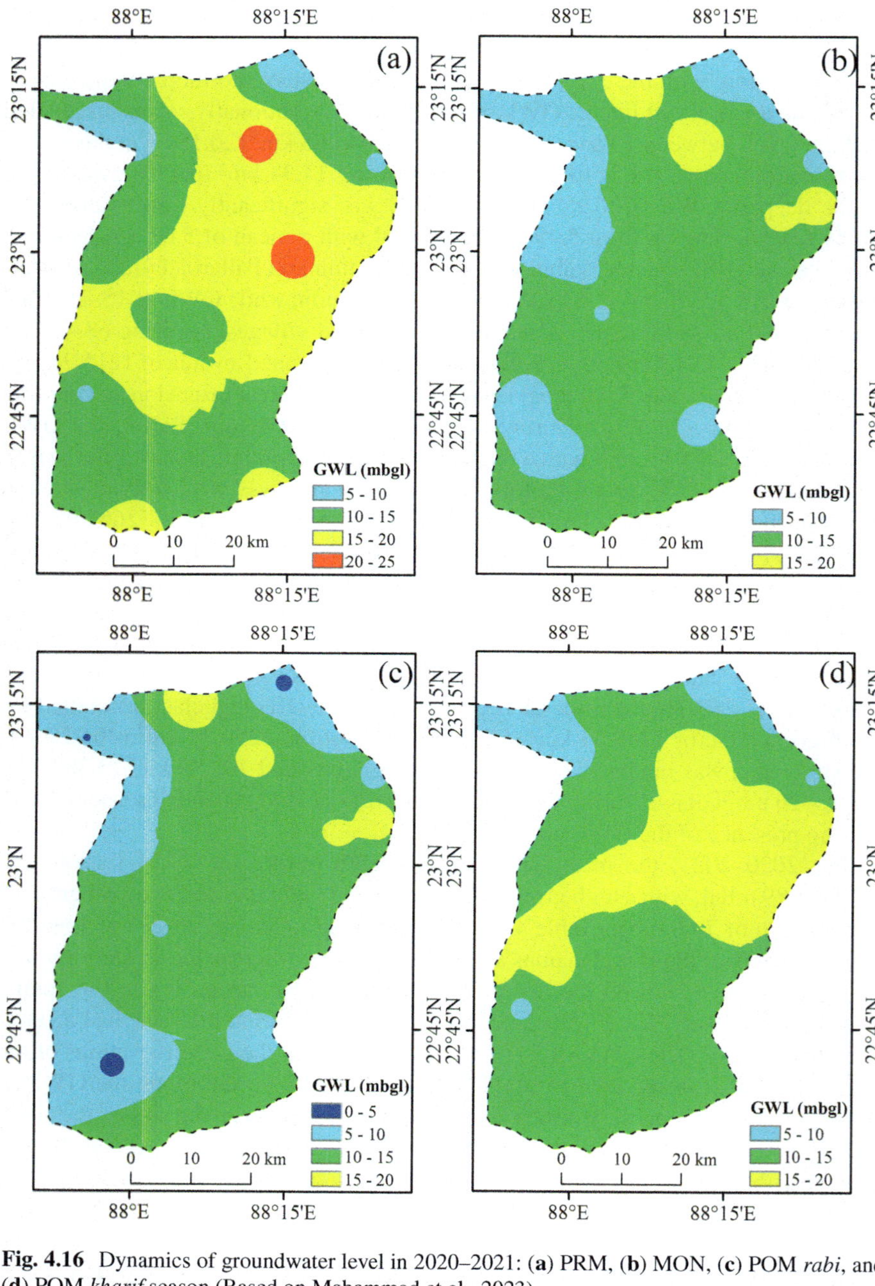

Fig. 4.16 Dynamics of groundwater level in 2020–2021: (**a**) PRM, (**b**) MON, (**c**) POM *rabi*, and (**d**) POM *kharif* season (Based on Mahammad et al., 2023)

Table 4.5 Spatial distribution of groundwater levels

Season	GWL (mbgl)	2013–2014 Area (km^2)	%	2020–2021 Area (km^2)	%	2025–2026 Area (km^2)	%
PRM	<5	1.77	0.06	0.00	0.00	0.00	0.00
	5–10	249.28	7.77	137.18	4.28	128.16	4.00
	10–15	2296.00	71.57	1410.28	43.96	1987.40	61.95
	15–20	646.54	20.15	1586.83	49.47	1038.26	32.37
	≥20	14.34	0.45	73.64	2.30	54.11	1.69
	Total	3207.94	100.00	3207.94	100.00	3207.94	100.00
MON	<5	52.74	1.64	0.00	0.00	0.00	0.00
	5–10	1870.96	58.32	672.92	20.98	157.90	4.92
	10–15	1215.74	37.90	2351.46	73.30	2737.37	85.33
	15–20	68.49	2.14	183.55	5.72	290.98	9.07
	≥20	0.00	0.00	0.00	0.00	21.68	0.68
	Total	3207.94	100.00	3207.94	100.00	3207.94	100.00
POM rabi	<5	180.79	5.64	20.13	0.63	1.61	0.05
	5–10	2667.97	83.17	852.29	26.57	433.23	13.50
	10–15	342.67	10.68	2201.37	68.62	2170.11	67.65
	15–20	16.50	0.51	134.15	4.18	584.84	18.23
	≥20	0.00	0.00	0.00	0.00	18.14	0.57
	Total	3207.94	100.00	3207.94	100.00	3207.94	100.00
POM kharif	<5	39.63	1.24	0.00	0.00	1.67	0.05
	5–10	1769.54	55.16	226.76	7.07	140.12	4.37
	10–15	1290.38	40.22	1976.49	61.61	1985.92	61.91
	15–20	108.39	3.38	1004.69	31.32	1064.35	33.18
	≥20	0.00	0.00	0.00	0.00	15.88	0.49
	Total	3207.94	100.00	3207.94	100.00	3207.94	100.00

Source: Mahammad et al., 2023

discovered in the northern region, which covered 183.55 km^2 (5.72%). During the POM *rabi* season, the GWL averaged 4.41 ± 0.51 mbgl, ranging from 3.9 to 18.88 mbgl. During this season, GWL of 0–5 mbgl covered 20.13 km^2 (0.63%), which is situated in the research area's southern and northernmost regions. The GWL of 5–10 is spread across the research area's western, southern, and northern borders, covering 852.29 km^2 (26.57%). Furthermore, the maximum research area of 2201.37 km^2 (68.62%) was determined to be covered by the 10–15 mbgl class. The northern portion of the GWL of 15–20 mbgl had an areal extension of 134.15 km^2 (4.18%). GWL varied from 6.73 to 19.6 mbgl with an average of 9.03 ± 2.3 mbgl throughout the POM *kharif* season. According to the geographical pattern, the GWL of the 5–10 mbgl class was spread in the northern portion of the research region, taking up a minimum of 226.76 km^2 (7.07%). In the research area's northern and southern regions, the minimum areal extension was 1976.49 km^2, or 61.61%, covered by the GWL of 10–15 mbgl. Additionally, it was discovered that class 15–20 mbgl was restricted to the central and eastern regions of the research area, covering an area of 1004.69 km^2 (31.32%). Furthermore, similar to 2013–2014, the MON, post-MON *rabi*, and post-MON *kharif* seasons did not yield the GWL class of ≥20 mbgl (Table 4.5).

The spatial distribution and change detection of GWL in the PRM, MON, POM *rabi*, and POM *kharif* seasons from 2013–2014 to 2020–2021 were shown (Fig. 4.17). The POM *kharif* season, PRM, MON, and POM *rabi* season saw the greatest changes in GWL, both positively and negatively. The northernmost point of the research region, Bara Dhamas, saw the greatest fall (~8 m) in GWL during the PRM season. A moderate drop in GWL was seen in the central portion of the study area. The research area's eastern region experienced minimal increases in GWL (~2 m). During the MON season, the research area's northern, southern, and westernmost regions had a decrease in GWL, while the eastern and western regions saw a rise in GWL. GWL showed an increasing tendency on a small scale (1 m) and a dropping trend in a few spots in the central, southern, and northern parts during the POM *rabi* season. Additionally, the POM *kharif* season had the greatest variation in the GWL, ranging from −6.95 to 9.22 m. The Parahati area showed an increase in GWL, while the remaining study area showed a fall in GWL up to ~9 m in depth.

Using an experimental variogram that shows geographical disparity in different periods, the spatial distributional pattern of the GWL in the research area was evaluated (Fig. 4.18a–h). PRM, MON, POM *rabi*, and POM *kharif* data for the assessment year 2013–2014 show a progressive change in the distributional pattern from a convex upward pattern from the pre to POM season to a linear trend (Fig. 4.18a–d).

This suggests that the boro rice cultivations' predominance of about equal water intake from various wells throughout the POM *kharif* season is the cause of the relative geographical equality. Though certain regions have continued to cultivate boro rice, there has been a minor decline in boro rice farming in recent years (2020–2021). Due to this, there is now regional inequality in the GWL withdrawal, particularly during the *kharif* season that follows the MON. During the course of the observations, this has caused the distribution to exhibit a convex upward pattern (Fig. 4.18e–h).

4.2.2.1 Model Findings and Performance of the Models

The best-fit model has been determined for each well (W1–W30) based on the Taylor diagram and reliability tests. The six models are the linear regressor, decision tree regressor, random forest regressor, k neighbour regressor, support vector regressor, and XGB regressor (Fig. 4.19). Every well's data was trained using 24 periods of the best-fit model, and eight periods were used for testing. The outcome demonstrates that the predictions for the maximum wells (W6, W7, W9, W14, W15, W16, W17, W23, W24, W25, W26, and W30) are comparable to the data from training and testing. That is, these wells suggest that the prognosis will be rather accurate. These wells (W1, W2, W27, and W24, for example) demonstrate that the prediction does not always match the training and testing datasets. Less future prediction reliability is implied by the difference in this regard. The prediction tool indicates that when temporal changes are taken into account, wells such as W2, W6, W7, W9,

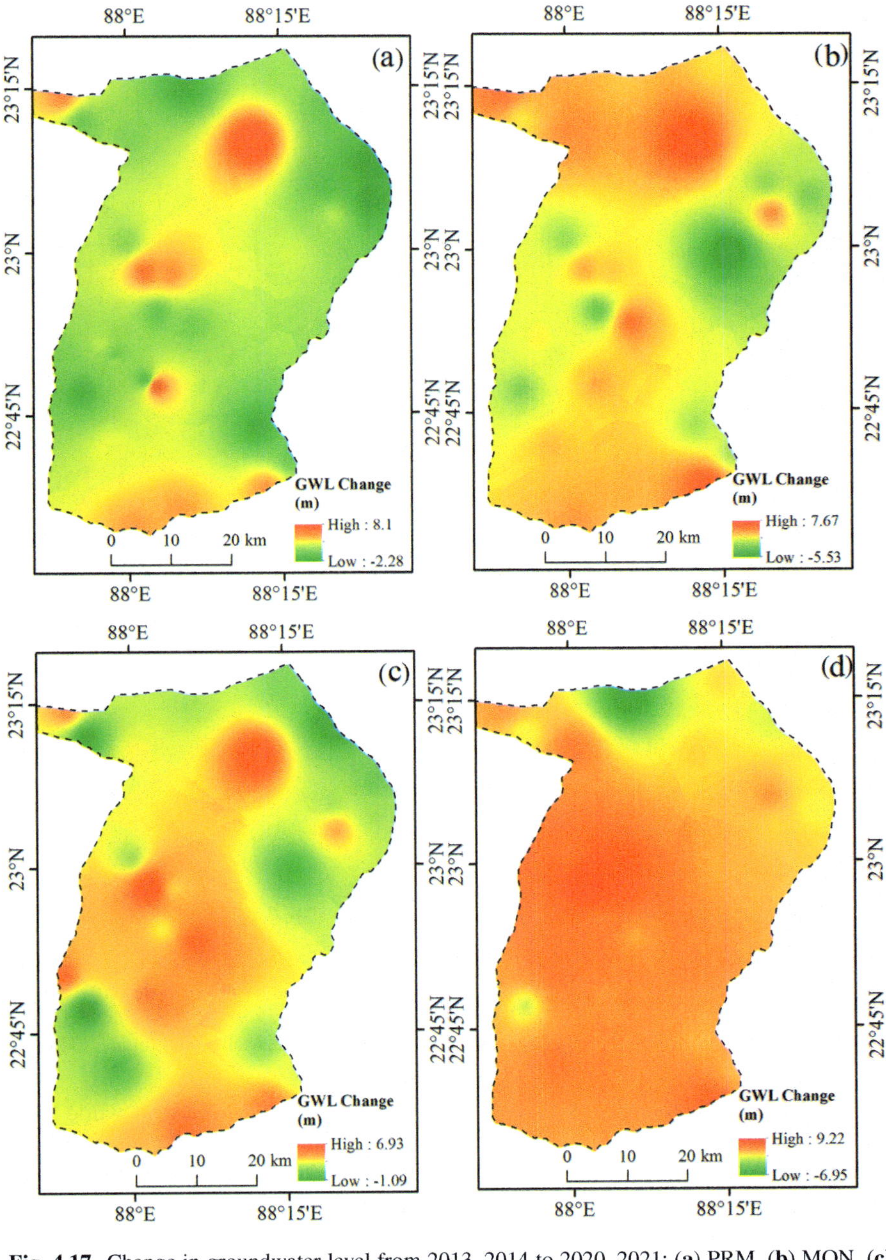

Fig. 4.17 Change in groundwater level from 2013–2014 to 2020–2021: (**a**) PRM, (**b**) MON, (**c**) POM *rabi*, and (**d**) POM *kharif* season (Source: Mahammad et al., 2023)

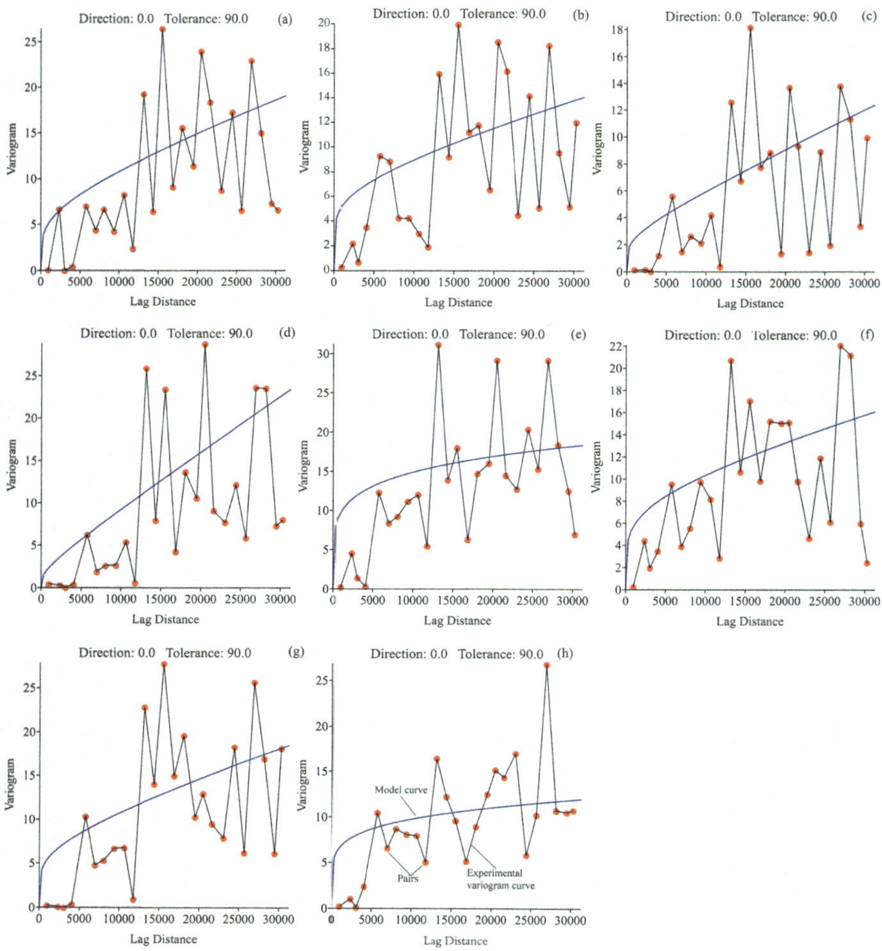

Fig. 4.18 Experimental variogram to show the spatial distributional pattern in the different seasons: (**a**) PRM 2013, (**b**) MON 2013, (**c**) POM *rabi 2013*, (**d**) POM *kharif 2014,* (**e**) PRM 2020, (**f**) MON 2020, (**g**) POM *rabi* 2020, and (**h**) POM *kharif* 2021 (Source: Mahammad et al., 2023)

W11, W14, W15, and W27 will exhibit an increasing trend in GWL, while wells W1, W22, W26, and W29 will exhibit a declining trend in GWL. But a lot of wells (such W3, W5, W8, W10, W11, and W24) exhibit erratic trends. Several regression models were used in this work to determine the valuable prediction of groundwater level. As a result, r was used to evaluate the models' performance, and during the training phase, RMSE was calculated using observed and predicted values (Table 4.6).

It is stated that in the case of RMSE, the model with a lower value signifies the accurate one, while a larger value of r indicates a more precise model (Fig. 4.20). The findings showed that, with r-values ranging from −0.72 to 0.9 and RMSE from

Fig. 4.19 Performance of the best fit model (Note: horizontal axes of the figures indicate the time span extending from 2013–2014 to 2020–2021; in a year (say 2013–2014), there are four seasons: the first point denotes PRM, second point for MON, third point for POM *rabi*, and fourth point for POM *kharif*) (Source: Mahammad et al., 2023)

4.73 to 0.52, the XGB regressor is the best model. The RF, LR, KN, SVR, and DT regressors also included r and RMSE values. With the largest RMSE, varying from 0.64 to 7.07, and the lowest r-value, ranging from −0.24 to 0.91, the DT regressor performed the worst out of all the sample wells. The Taylor diagram has also been

Table 4.6 Performance of the models (Source: Mahammad et al., 2023)

Well	Correlation coefficient (*r*)						Root mean square error (RMSE)						Best fit
ID	DT	RF	LR	KN	SVR	XGB	DT	RF	LR	KN	SVR	XGB	model
W1	0.24	0.26	−0.12	0.59	0.19	0.57	2.98	2.73	3.36	2.18	2.69	3.04	KN
W2	−0.20	0.12	0.68	−0.28	0.24	−0.06	3.10	3.08	1.54	4.11	3.46	3.25	LR
W3	0.68	0.63	0.77	0.67	0.68	0.62	2.68	2.36	2.39	2.62	2.71	2.81	RF
W4	0.82	0.92	0.54	0.91	0.74	0.88	1.95	1.75	0.62	1.80	1.86	1.66	LR
W5	0.28	0.24	−0.20	−0.07	0.09	0.82	1.79	1.42	1.30	1.73	1.34	1.04	XGB
W6	0.91	0.87	0.55	0.92	0.79	0.41	1.37	1.78	1.95	0.98	1.78	1.85	KN
W7	0.69	0.75	0.50	0.25	0.45	0.90	1.16	0.82	1.07	1.20	1.03	0.52	XGB
W8	0.58	0.01	0.37	0.66	0.64	−0.19	2.36	2.42	2.45	2.26	2.41	2.49	KN
W9	0.76	0.45	0.58	0.43	0.85	0.75	1.46	2.04	1.97	2.88	1.16	1.74	SVR
W10	0.56	0.75	0.11	0.82	0.76	0.42	0.86	0.53	3.44	0.56	0.66	1.02	KN
W11	0.03	0.28	0.57	0.28	0.23	0.20	1.92	1.48	6.13	1.52	1.54	1.90	RF
W12	0.48	0.67	0.56	0.30	0.76	0.60	3.92	1.50	2.33	4.34	2.47	2.00	RF
W13	0.34	0.75	0.92	0.82	0.84	0.86	2.82	2.47	1.43	2.04	2.55	2.21	LR
W14	0.69	0.70	0.91	0.77	0.81	0.78	3.32	3.07	1.99	3.04	3.40	2.70	LR
W15	0.81	0.82	0.92	0.86	0.77	0.79	1.88	1.63	0.88	1.53	2.00	1.70	LR
W16	0.11	0.57	0.60	0.45	0.32	0.82	2.53	1.49	2.03	2.20	2.03	1.23	XGB
W17	0.68	0.70	0.57	0.76	0.62	0.79	2.23	1.89	1.89	1.76	1.72	1.43	XGB
W18	0.02	0.03	0.10	−0.13	0.24	−0.72	7.07	2.74	6.80	3.72	2.07	4.12	SVR
W19	0.51	0.59	0.60	0.33	0.57	0.81	1.77	1.90	1.43	2.59	1.54	0.87	XGB
W20	0.28	0.51	0.39	0.53	0.37	0.69	3.36	3.36	4.20	3.40	2.83	2.61	XGB
W21	0.90	0.68	0.69	0.68	0.57	0.25	2.68	4.06	3.90	4.08	4.66	4.73	LR
W22	−0.16	0.02	0.81	0.56	0.76	0.70	1.29	1.28	1.19	1.32	1.29	1.04	XGB
W23	0.59	0.90	0.88	0.81	0.89	0.72	0.64	1.07	1.04	1.42	1.62	1.41	LR
W24	0.75	0.82	0.30	0.08	0.02	0.57	1.81	1.69	3.28	3.17	2.55	2.32	RF
W25	0.81	0.60	0.78	0.19	0.54	0.39	2.01	1.93	1.37	3.02	1.92	2.39	SVR
W26	0.72	0.73	0.52	0.75	0.82	0.86	1.50	1.56	4.63	1.71	1.54	1.16	XGB
W27	0.43	0.68	0.54	0.64	0.36	0.42	1.44	0.89	1.25	1.47	1.13	1.36	RF
W28	0.73	0.68	0.46	0.66	0.68	0.51	4.35	3.35	2.61	2.74	3.52	2.58	XGB
W29	−0.10	0.30	0.50	0.20	0.70	0.41	1.21	1.13	1.07	1.77	0.89	1.03	XGB
W30	−0.24	0.76	0.66	0.90	0.87	0.61	4.22	1.95	2.39	1.36	2.24	2.47	RF

used to examine the validity and reliability of several machine learning methods for various wells ($N = 30$). A Taylor diagram is a graphical representation created using RMSE, SD, and r. The various models fitted to the Taylor diagram are evaluated in relation to the observed values of RMSE, SD, and r. A model with a maximum r, a minimal SD, and an RMSE is the best fit. This finds the model that fits the data closest to the observed line. In this manner, as the 30 Taylor diagrams demonstrate (Fig. 4.20), the XGB regressor emerges as the most dependable model for prediction. This suggests that most of the wells' water level distribution is erratic.

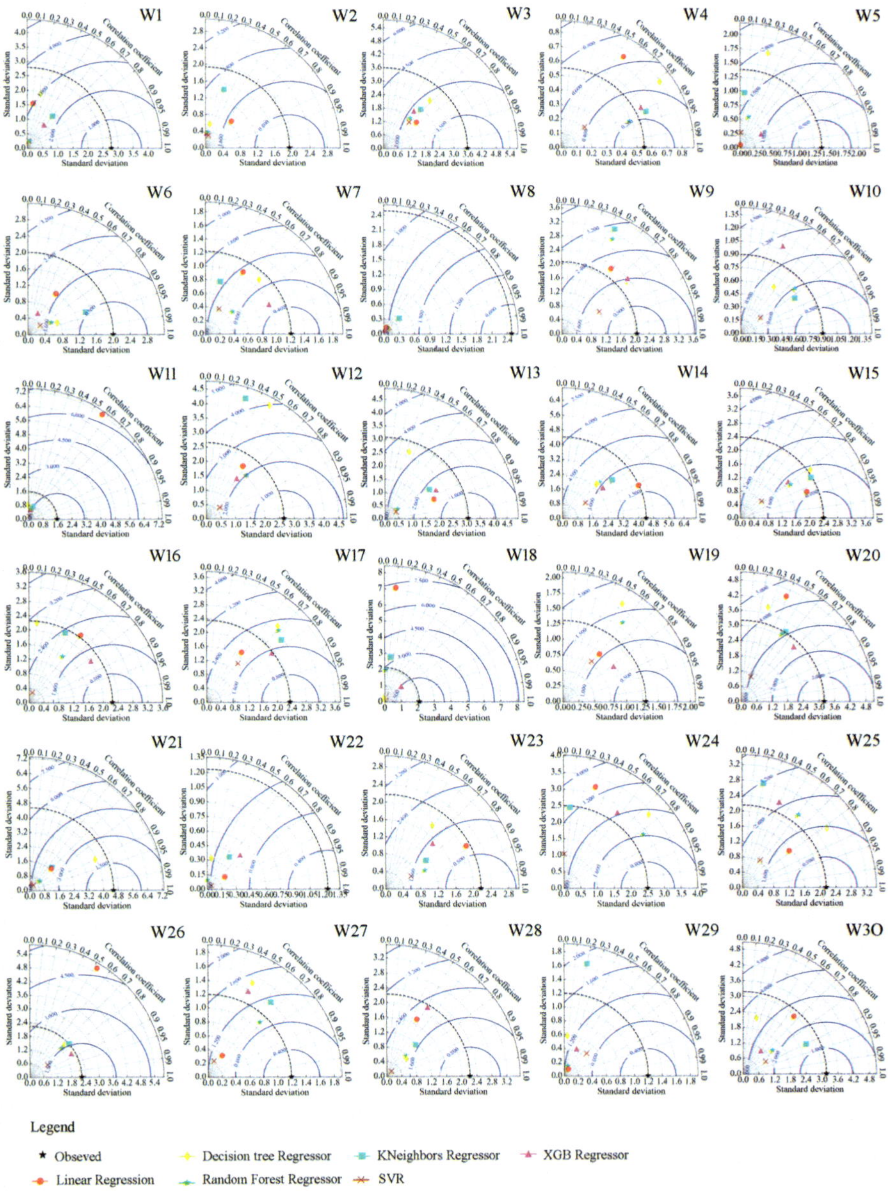

Fig. 4.20 Taylor diagram showing the accuracy of the different models based on RMSE, *r*, and SD (Source: Mahammad et al., 2023)

4.2.2.2 Future Predictions

Regression modelling has been used to estimate the distribution of the GWL for 2025–2026; this model has been shown to be the best fit in terms of accuracy evaluation. According to the simulated GWL results, during all seasons—PRM, MON, POM *rabi*, and POM *kharif*—GWL of less than 10 mbgl drastically dropped, whereas GWL of more than 15 mbgl increased more considerably (Fig. 4.21). The predicted GWL's geographical distribution showed that the GWL class of 0–5 mbgl is absent during the PRM season, whereas the GWL class of 5–10 mbgl is found to extend for 128.16 km^2 (4%) and decrease by about 9 km^2 from 2020–2021. This season's maximum GWL of 10–15 mbgl, or 1987.4 km^2, was reached (61.95%).

The research area's GWL of 15–20 and >20 mbgl, which are primarily in the northern and central regions, respectively, covers 32.37 and 1.69% of the study area. The majority of the study region during the MON season (85.33%) is covered by GWL of 10–15 mbgl, with the remaining portion being covered by GWL of 5–10, 15–20, and >20, which is limited to a very small number of places. In the northernmost portion of the research region during the POM *rabi* season, there was GWL of 0–5 and 5–10 mbgl in tiny areas of extension, making up 0.05% and 13.5% of the total. The research area was covered by the class of 10–15 mbgl, accounting for 67.65%, while the class of 15–20 mbgl accounted for 18.23%. Furthermore, the forecast year, which covers 18.14 km^2, has a GWL of >20 mbgl, in contrast to the POM *rabi* seasons of 2013–2014 and 2020–2021. During the POM *kharif* season the GWL 5–10 mbgl is situated at Simlon located in the northernmost part of the study area. The GWL of 5–10 mbgl is situated in the exact region, and the maximum extension in the 1985.92 km^2 (61.91%) study area is covered by 10–15 mbgl. Within the study area, the class of 15–20 mbgl covers 1064.35 km^2, primarily in the western and middle-eastern regions. In addition, a 20–25 mbgl class was discovered in the expected year's POM *kharif* season compared to the previous year. A change analysis from 2020 to 2021 was conducted for the anticipated year 2025–2026, and the results showed that GWL declined by up to ~16 m (Fig. 4.22). Additionally, it was noted that during all seasons, the research area's westernmost region, which includes Barddhaman town, was the primary location for the drop in GWL. Furthermore, there have been a few spots where GWL has increased, although they have only done so up to about 3 meters.

4.3 Assessment of Surface Water Quality for Drinking and Irrigation Purposes

4.3.1 Assessment of Drinking Water Quality

The average surface water quality is 70 ± 11.15, with values from 51.63 to 98.64. As a result, surface water with a WQI of 50–100 is regarded as good water (Wagh et al., 2017). The Water Quality Index (WQI) has been divided into five classes—excellent

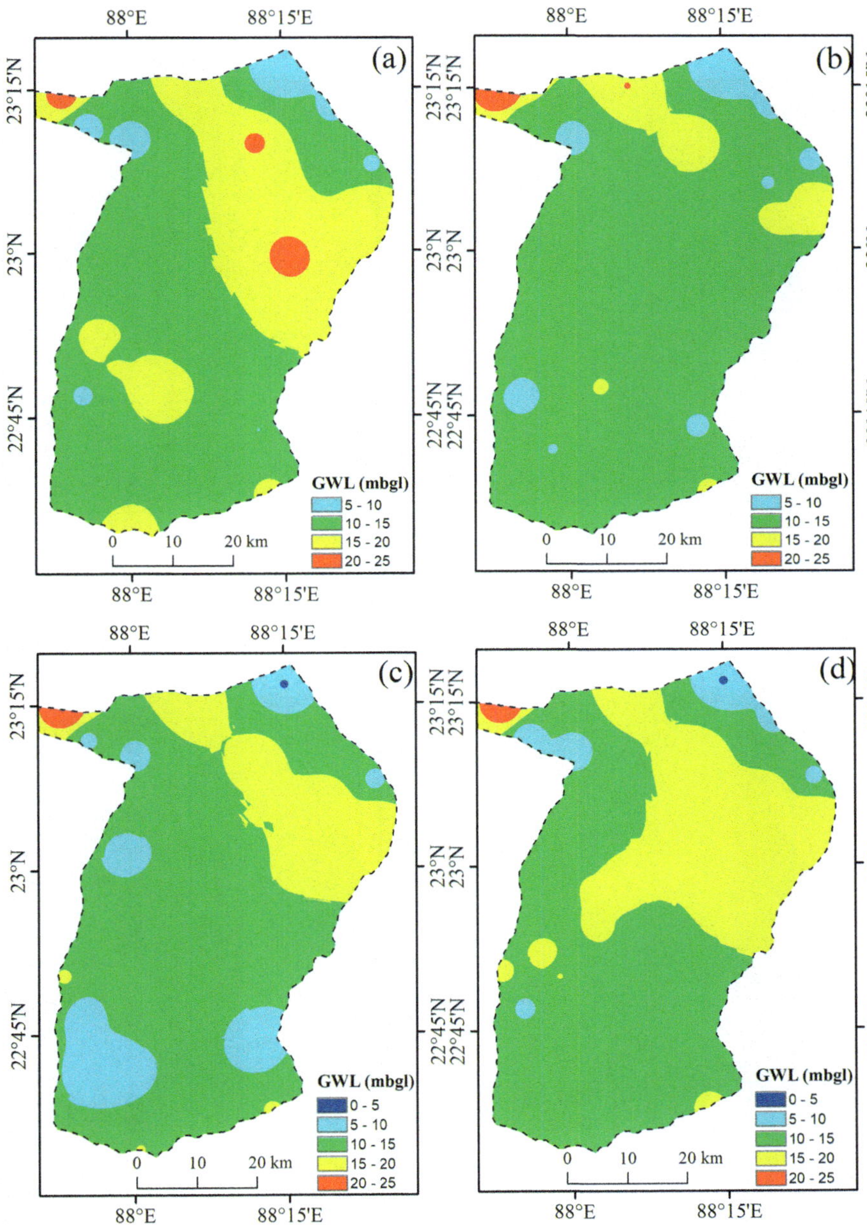

Fig. 4.21 Predicted groundwater level of 2025–2026: (a) PRM, (b) monsoon, (c) POM *rabi*, and (d) POM *kharif* season (Based on Mahammad et al., 2023)

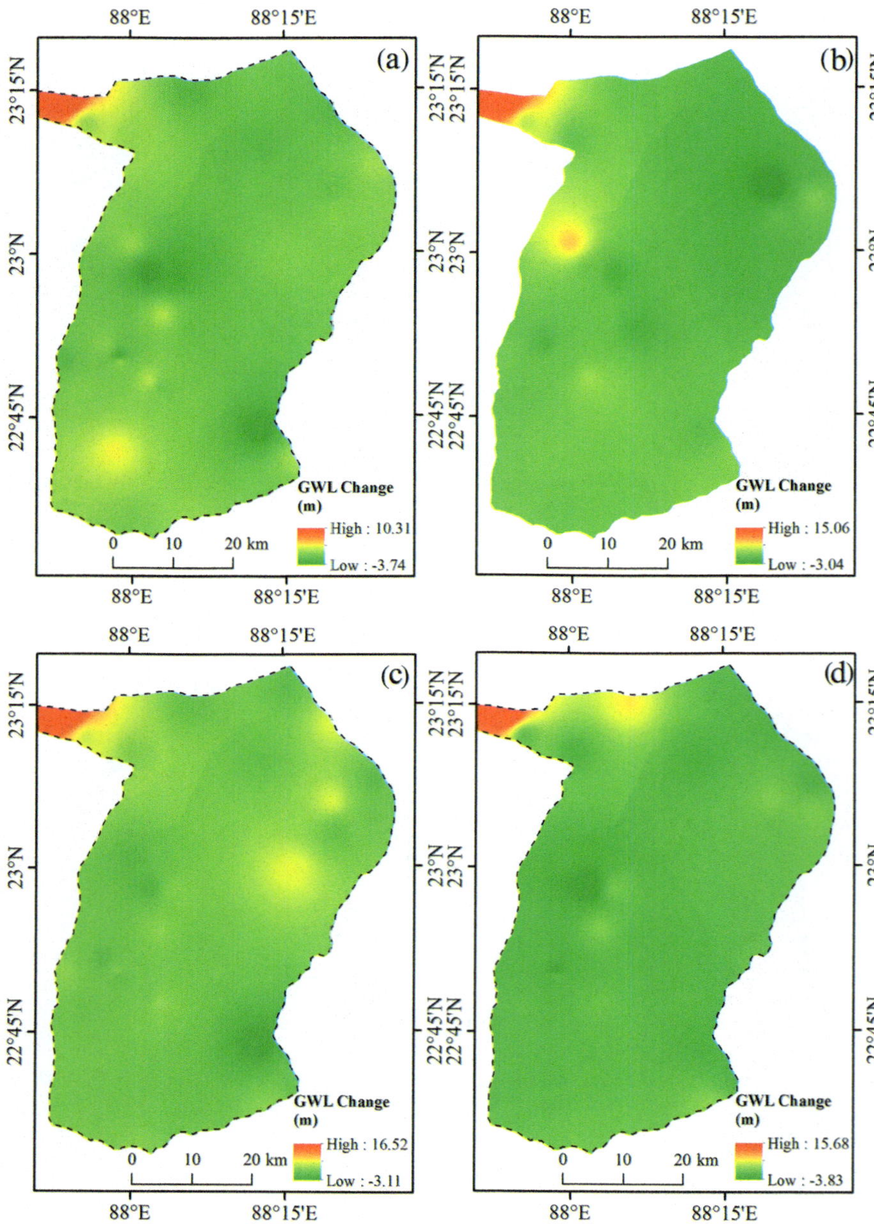

Fig. 4.22 Change in groundwater level from 2020–2021 to 2025–2026: (**a**) PRM, (**b**) monsoon, (**c**) POM *rabi*, and (**d**) POM *kharif* season (Source: Mahammad et al., 2023)

(50–60), good (60–70), moderate (70–80), poor (80–90), and very poor (90–100)—in order to illustrate regional variance. The findings showed that 19.44% (7) of the sample surface water falls under the excellent class of WQI values, which indicates that the water quality in the research area is generally the best.

Furthermore, 33.33% (11) of the sample water falls into the good class of WQI value. Furthermore, 30.56% of the sample surface water belongs to the moderate class of WQI. Furthermore, of the sample water in the study area, 11.11% and 5.55% fall into the poor and very poor class of WQI. Furthermore, the research area's northern region is home to the very low and poor class of water quality, according to the geographic distribution of surface water quality (Fig. 4.23). The moderate class extends the study area to its greatest extent. Additionally, the research area's western, eastern, and central regions are home to a large concentration of circular patches representing the high and very high class of WQI. The western

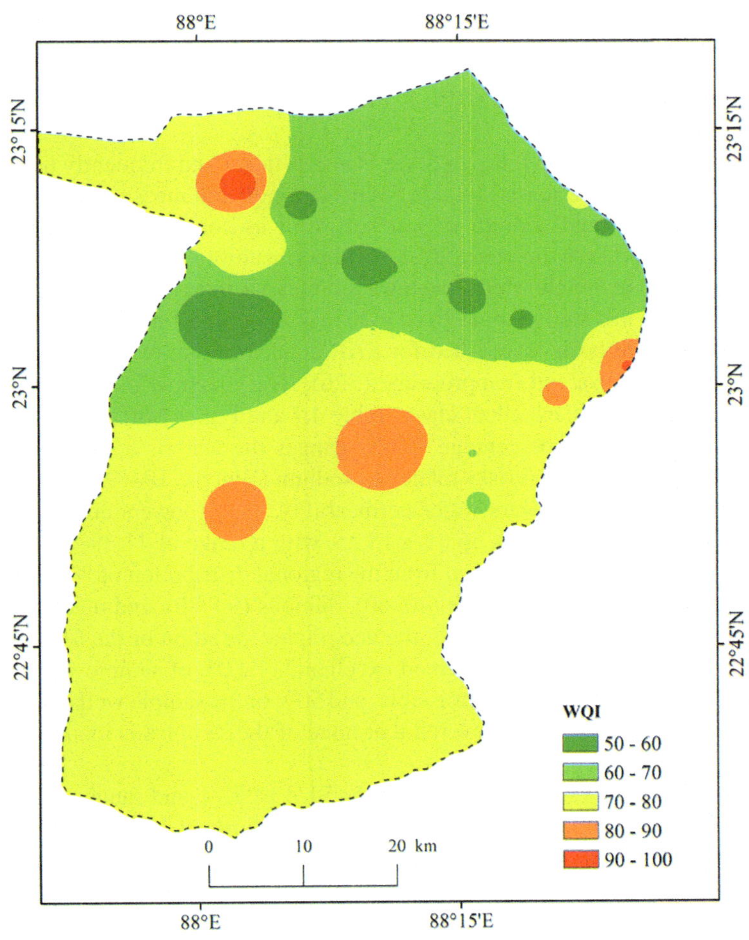

Fig. 4.23 Spatial distribution of Surface Water Quality Index for drinking purposes

region, close to Barddhaman town, has a higher WQI value, suggesting the presence of pollutants from the surrounding metropolitan area. Additionally, it has been discovered that the Kunti Nadi's greater WQI value is restricted there as a result of industrial effluent discharge (Panigrahi & Pattnaik, 2020).

4.3.2 *Assessment of Irrigation Water Quality*

4.3.2.1 Irrigation Hazards

Electrical conductivity (EC) is a measure of dissolved ions in water and is typically used to estimate the salinity of the water (Adimalla et al., 2020). Thus, the salt content of the water measured is enriched by a high EC concentration. According to FAO (1985), there are four types of EC concentration in this study: none (<700 μS/cm), weak to moderate (700–3000 μS/cm), and severe (>3000 μS/cm). According to the spatial distribution, 8.33% of the water sample is classified as mild to moderately contaminated, while 91.67% of the sample is deemed much less polluted (Table 4.7 and Fig. 4.24a).

The sodium absorption ratio, or SAR, is a significant and frequently used indicator to assess whether water is suitable for irrigation. It can measure the ratio of calcium, magnesium, and sodium in water as well as convey the extent of cation exchange in the soil (Aydin et al., 2021). Soil permeability is decreased and salinity is increased due to the elevated concentration of SAR in irrigation water. The SAR concentration in this study ranges from 0.06 to 4.53, with an average of 2.25 ± 1.15. Three grades have been established for SAR concentration: excellent (<10), good (10–18), fair (18–26), and inappropriate (>26). Every sample in this investigation was judged to be in the excellent class (Table 4.7 and Fig. 4.24b).

The sodium content percentage (%Na) gauges the water's appropriateness for agriculture and its potential risks related to sodium (Wilcox, 1948). A high quantity of Na in surface water lowers water permeability, which lowers crop yields. This study's mean value of %Na is 56.01 ± 15.25, with a range of 11.39–76.47%. Four classifications have been identified from the regional distribution of %Na: excellent (0–20), good (20–40), permissible (40–60), dubious (60–80), and unsuitable (>80) (Fig. 4.24c). According to Table 4.7, the geographic variation of the %Na indicates that 2.78% of the samples are deemed excellent, 11.11% of sample wells account for good, 36.11% are deemed permissible, and 50% of the sample wells are observed as poor. Furthermore, it is discovered that none of the samples is inappropriate for irrigation.

This study's SSP value has a mean of 40.54 ± 14.92% and ranges from 2.36 to 62.15%. The concentration of SSP in this investigation has been divided into two categories: appropriate (<50) and unsuitable (>50). According to the spatial variation of the SSP, 66.67% of the samples are deemed appropriate for irrigation, while 33.33% are deemed unsuitable (Table 4.7 and Fig. 4.24d).

Table 4.7 Parameter-specific water suitability classes for irrigation and the distribution of surface water samples

Parameters	Range	Water category	No. of wells	Percentage of wells
pH	<6.5	Unsuitable	0	0
	6.5–8.5	Suitable	36	100
	>8.5	Unsuitable	0	0
Cl^- (mg/L)	<354	Suitable	36	100
	>354	Unsuitable	0	0
NO_3^- (mg/L)	<30	Suitable	36	100
	>30	Unsuitable	0	0
EC (μS/cm)	<700	None	33	91.67
	700–3000	Slight–moderate	3	8.33
	>3000	Severe	0	0
SAR	0–10	Excellent	36	100
	10–18	Good	0	0
	18–26	Fair	0	0
	>26	Unsuitable	0	0
%Na	0–20	Excellent	1	2.78
	20–40	Good	4	11.11
	40–60	Permissible	13	36.11
	60–80	Poor	18	50
	>80	Unsuitable	0	0
SSP (%)	<50	Good	24	66.67
	>50	Unsuitable	12	33.33
RSC (meq/L)	<1.25	Good	8	22.22
	1.25–2.5	Moderate	4	11.11
	>2.5	Bad	24	66.67
MH (%)	<50	Good	28	77.77
	>50	Unsuitable	8	22.22
KR	<1	Suitable	17	47.22
	>1	Unsuitable	19	52.78
PI (%)	<80	Good	14	38.89
	80–100	Moderate	21	58.33
	100–120	Poor	1	2.78
	>120	Unsuitable	0	0
EIWQI	25–50	Excellent	7	19.44
	50–75	Good	17	47.22
	75–100	Moderate	11	30.55
	100–125	Poor	1	2.78

Source: Computed by the authors, 2021

Based on EC and salt concentration, surface water samples were divided into five categories in this study using the Wilcox diagram: (1) excellent to good, (2) good to permissible, (3) permissible to questionable, (4) doubtful to unsuitable, and (5) unsuitable (Fig. 4.25a). About 80% of the water samples in this study fit into the

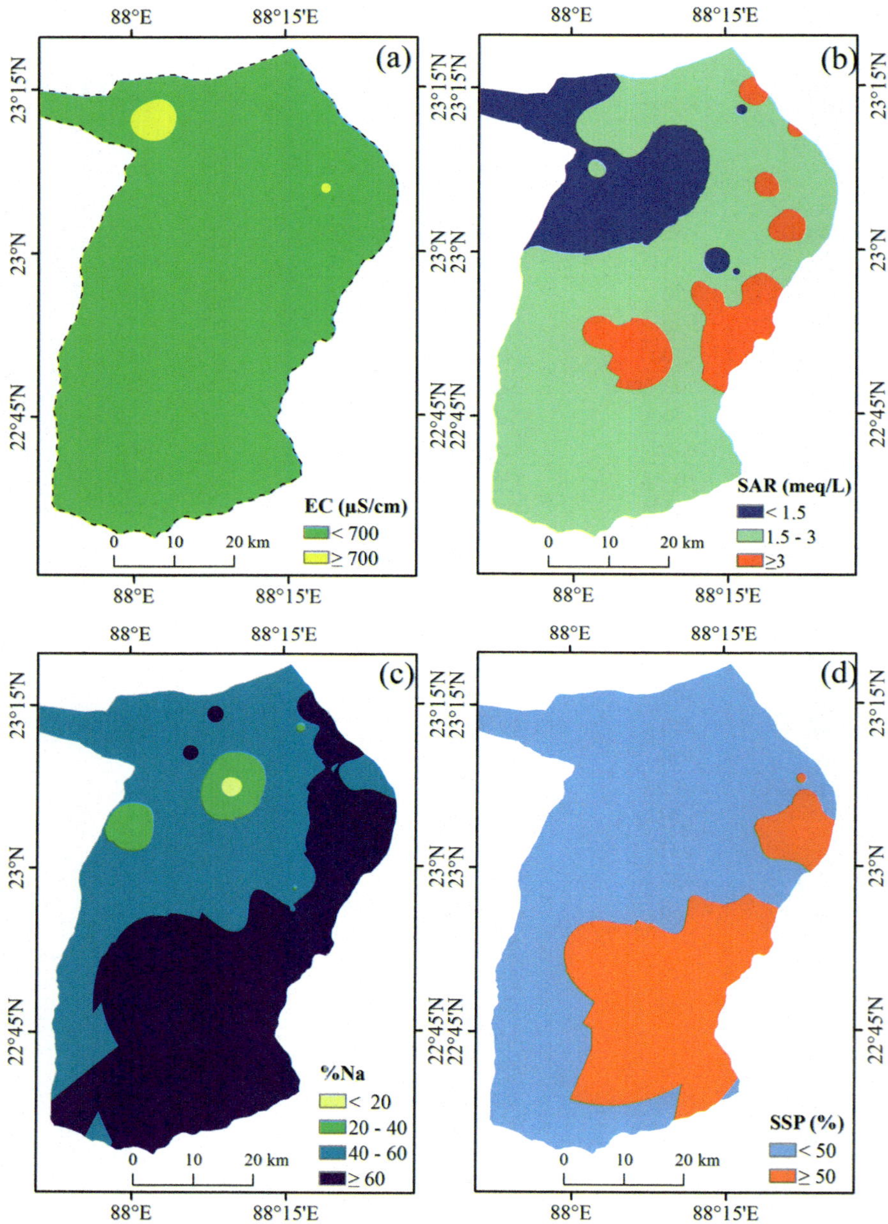

Fig. 4.24 Spatial distribution of irrigation water quality parameters and indices of surface water samples: (**a**) EC, (**b**) SAR, (**c**) % Na, and (**d**) SSP

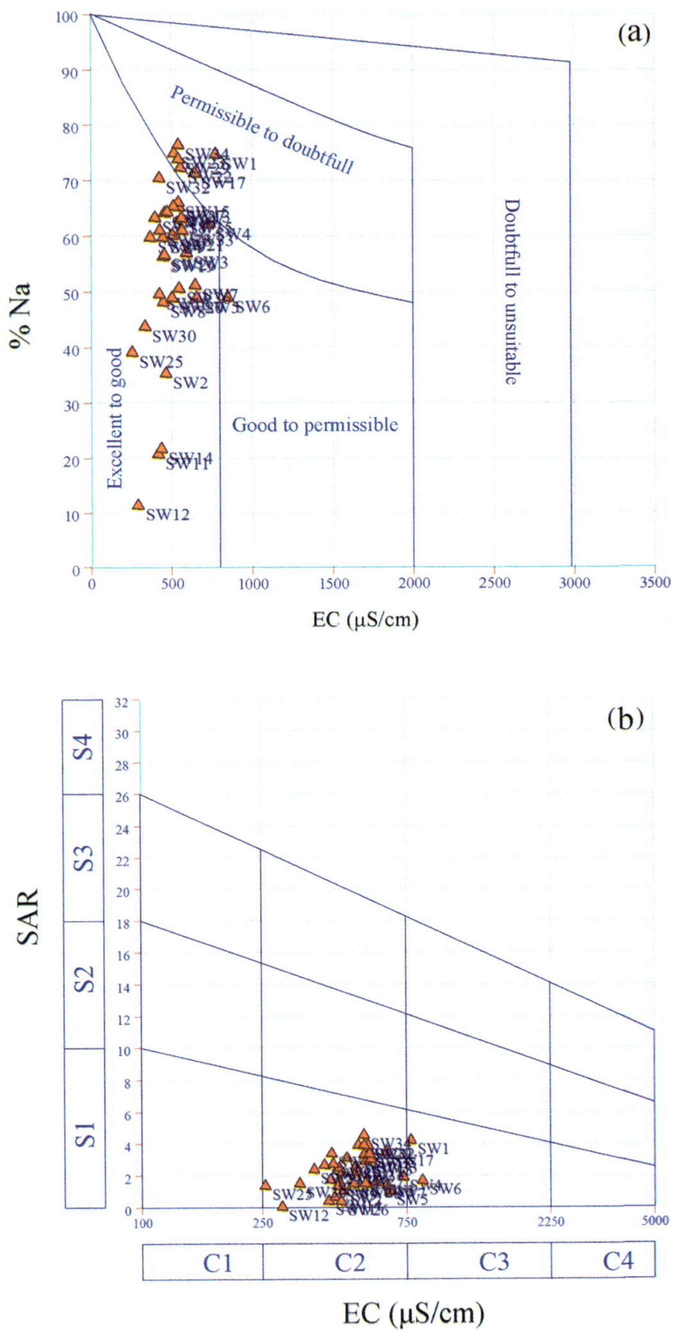

Fig. 4.25 Classification of surface water samples for irrigation: (**a**) Wilcox and (**b**) USSL diagram

excellent to good class for irrigation purposes, according to the Wilcox diagram. Furthermore, three percent of the water samples are suitable for agricultural purposes. Additionally, about 17% of the samples are shown to be acceptable for the dubious class.

Based on electrical conductivity and SAR value, irrigation water is categorized using the USSL method. The classification of the water into classes denoting salinity hazards (C1, C2, C3, and C4) and sodium hazards (S1, S2, S3, and S4) is revealed. Only one water sample is seen to fall into the C1 class on the USSL diagram (Fig. 4.25b), which indicates low salinity water that is appropriate for a wide variety of crops on all types of soils. Additionally, 92% of the samples fall into the C2 class, indicating that the water is in medium salinity level and suitable for irrigating crops that can withstand high salt levels (Balasubramani et al., 2020). Moreover, C3, which makes up less than 5% of the samples, is regarded as a high salinity water group. Additionally, it demonstrated that not a single sample of surface water belongs to the high salinity water (C4) category. Furthermore, every surface water sample belongs to the S1 class, which comprises low-sodium water. For the majority of crops, this water is suitable because its detrimental sodium exchangeable content is low (Balasubramani et al., 2020).

When determining whether sodium cations are present in water, residual sodium concentration, or RSC, is a helpful indicator. For this investigation, the RSC concentration has been measured at each sample location and has been found to vary from −2.37 to 14.6 meq/L. The average RSC concentration was determined to be 4.66 ± 4.18 meq/L. Three categories have been established for RSC concentration classification: excellent (<1.25), moderate (1.25–2.5), and bad (>2.5). According to the data, 22.22% of the surface water samples fall into the good category, while 11.11% fall into the intermediate category. Furthermore, 66.67% of the surface water samples fall into the unsuitable category for irrigation in the research area (Table 4.7 and Fig. 4.26a).

Magnesium hazard (MH) in the research area ranges from 13.14% to 80.55%, with an average of 39.22 ± 15.26%. According to Narasaiah and Rao (2021a, b), the MH requirements for irrigation should be 50%. Consequently, water with an MH content more than 50 is deemed unfit for irrigation. Consequently, according to the MH content, 22.23% of the sample water is rated as good, while 77.77% of the sample well is deemed unfit for irrigation (Fig. 4.26b).

Because irrigation water's high sodium content retards plant growth, the Kelley's ratio (KR) is a useful tool for assessing the appropriateness of irrigation water (Kelley et al., 1940). Water having a KR value above 50 is considered unfit for irrigation and vice versa. The KR content in this study varies from 0.03 to 2.64, with an average of 39.22 ± 15.26. It is determined that 52.78% of the study's sample water is not suitable for irrigation since it surpasses the threshold value of 1 (Fig. 4.26c and Table 4.7).

The potentiality index (PI) value in this study has a mean of 80.94 ± 13.6 and varies from 46.73 to 100.93. Doneen (1964) distinguished four groups of PI concentrations: excellent (<80), moderate (80–100), poor (100–120), and unsuitable (>120) for irrigation. Based on PI value concentration, three classifications have

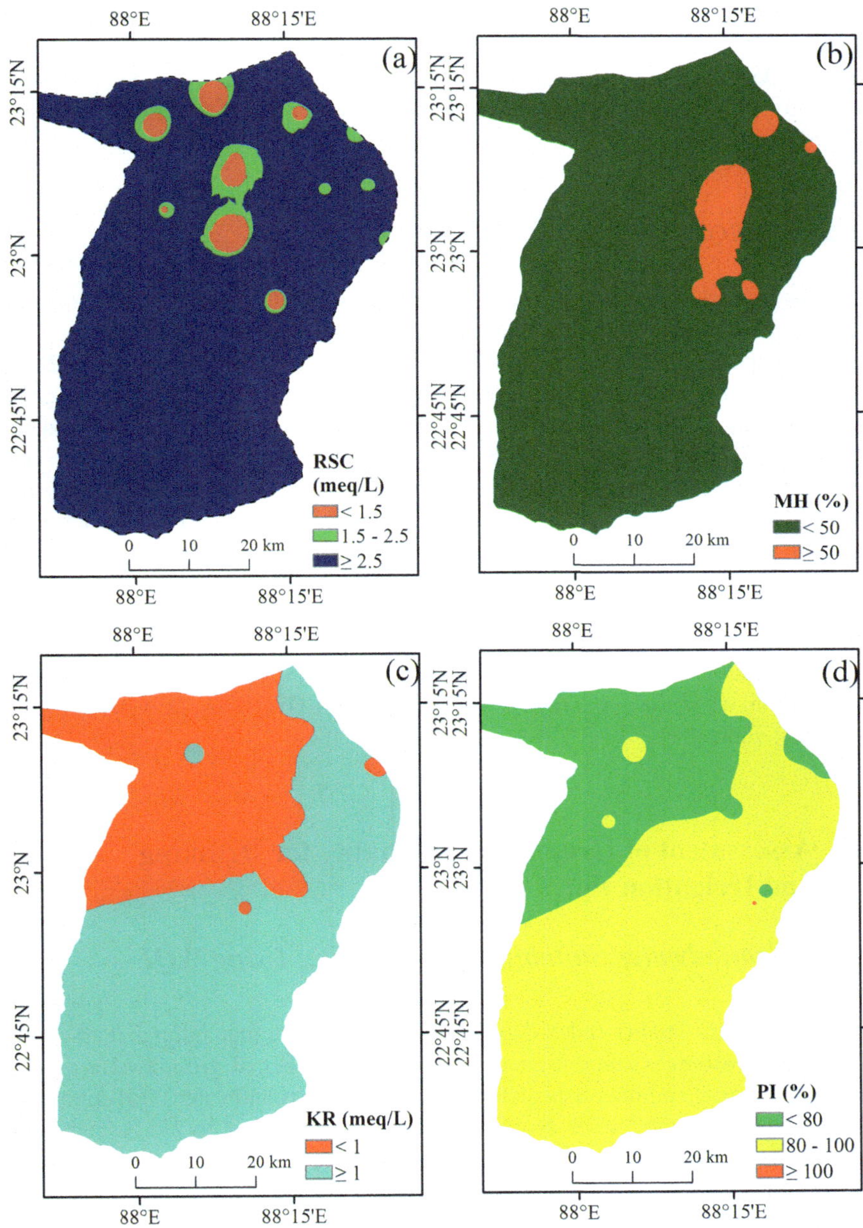

Fig. 4.26 Spatial distribution of irrigation water quality indices of surface water samples: (**a**) RSC, (**b**) MH, (**c**) KR, and (**d**) PI

been identified: good, midrange, and poor. The PI value distribution indicates that 38.89% of the sample is deemed good for irrigation use, 58.33% of the sample water is deemed moderate, and 2.78% of the sample water is deemed poor (Fig. 4.26d). Furthermore, it has been determined that all of the sample water is appropriate for irrigation.

4.3.2.2 EIWQI

The entropy-based irrigation water quality index (EIWQI), which yielded a single irrigation suitability score, was used to quantify the level of water quality. This study's computed EIWQI score had a mean of 71.94 ± 23.77 and ranged from 26.48 to 118.57. Higher scores on the EIWQI indicate lower irrigation water quality. The index assesses irrigation water quality. In this study, the EIWQI has been categorized into four groups: excellent (25–50), good (50–75), moderate (75–100), and poor (>100) (refer to Table 4.7). According to the results, 19.44% of the samples are considered excellent, and just 2.78% are not appropriate for irrigation. Furthermore, moderate and poor water quality classes are assigned classifications of 47.22% and 30.55%, respectively. Consequently, it also showed that the vast majority of samples in the study area contain water that is suitable for irrigation (good to outstanding). Furthermore, the EIWQI's spatial distribution reveals that the research area's eastern and southern regions are concentrated in areas with lower surface water quality for irrigation, while the upper region has the superior EIWQI surface water category (Fig. 4.27; Table 4.8).

4.4 Assessment of Groundwater Quality for Drinking and Irrigation Purposes

4.4.1 Groundwater Suitability for Drinking Using WQI

The Groundwater Quality Index (WQI) is a useful tool for managing and assessing groundwater resources since it condenses a large number of groundwater quality factors into a single number using a basic manner. The groundwater WQI comprises an average value of 62.40 ± 26.46 and varies from 23 to 126.23. Nonetheless, according to Hamlat and Guidoum (2018) and Minh et al. (2019), the WQI has been categorized into excellent water (<50), acceptable water (50–100), bad water (100–200), extremely poor water (200–300), and unfit for human consumption (>300). The study's WQI has been divided into four groups: <50, 50–75, 75–100, and > 100 (Fig. 4.28).

The water has high quality if the WQI is less than 50. An index of 50–75 suggests good water quality, an index between 75 and 100 indicates poor water quality, and an index more than 100 indicates extremely poor water quality for drinking

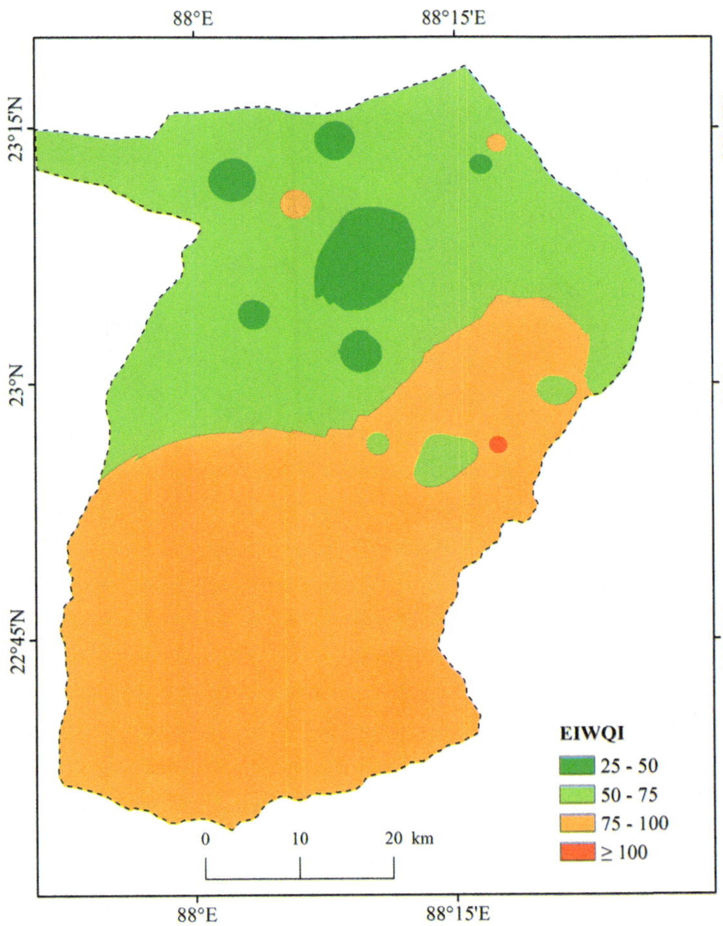

Fig. 4.27 Spatial distribution of EIWQI

Table 4.8 Acceptable limits and relative weights of the water quality parameters for irrigation purposes

Parameters	Acceptable limits	Relative weights
pH	6.5–8.5	0.092
Cl^- (mg/L)	354	0.094
NO_3^- (mg/L)	30	0.082
EC (μs/cm)	3000	0.097
SAR	26	0.090
%Na	80	0.097
SSP (%)	50	0.096
RSC (meq/L)	2.5	0.086
MH (%)	50	0.085
KR	1	0.086
PI (%)	100	0.095

Source: Computed by the authors, 2021

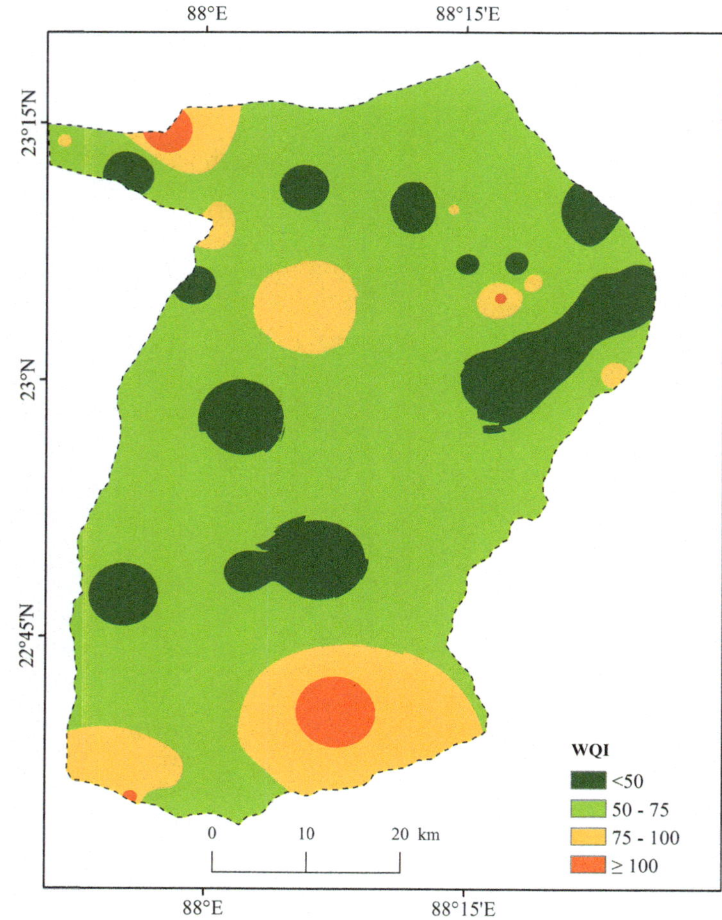

Fig. 4.28 Spatial variation in WQI

purposes in this research region. The outcome reveals that 4 out of the 17 sample wells, or 50% of the total, are of extremely low quality, while the remaining 17 wells, or 11.76%, are of exceptional quality. The findings also indicate that, in this study region, 17.65% of the samples (six wells) are classified as very excellent quality and 20.59% of the samples (seven wells) as poor water quality for drinking (Fig. 4.28). Very poor-quality water can be found in the research area's northwest, south, and east, according to the WQI's spatial distribution. The southern portion of the research region, northwest of Barddhaman town, is where the poor water is concentrated. The western, northeastern, and central regions of the research area are home to the wells with exceptional water quality. Wells with high water quality can also be found across the remaining research area.

4.4.2 *Nature of Irrigation Quality*

4.4.2.1 Nature of Irrigation Hazards

According to Bozdag (2014), the EC is a measure of the dissolved ions in water and typically indicates the salt concentration in the water. Thus, the salt content of the groundwater analysed is enriched by a high EC concentration (Adimalla et al., 2018). According to Table 4.9, this investigation indicates that the average concentration of EC is 760.53 ± 449.85 μS/cm, with a range of 184–2002 μS/cm. Three kinds of EC concentrations have been identified: semi-tolerant (1500–3000), tolerant (>3000), and sensitive (<1500). According to the geographical variation of the EC, 97.18% of the water sample is deemed sensitive (Fig. 4.29a), while 8.82% of the samples belong to the semi-tolerant class (Table 4.10). Nevertheless, it was discovered that none of the groundwater samples belonged to the tolerant class.

Since it assesses the risk of alkali and/or sodium to crops, the SAR is the most important and commonly recognized quality characteristic of the water for irrigation (Subramani et al., 2005). Table 4.9 shows that the SAR values for water samples range from 0.49 to 14.63, with an average of 2.64 ± 2.46. Four classes can be seen in the SAR's spatial distribution as well. Less than 10 SAR concentrations in water samples indicate an excellent water type; 10–18 SAR is regarded as good; 18–26 SAR is regarded as fair; and more than 26 SAR is regarded as unsuitable for irrigation (USSL, 1954). With the exception of one sample (GWQ26), all of the samples in this investigation are outstanding (Fig. 4.29b). As a result, 2.94% of the water samples are good for irrigation and 97.06% of the water samples are excellent for irrigation (Table 4.10).

In addition, the percentage Na is a crucial indicator for assessing whether groundwater is suitable for irrigation (Wilcox, 1948). Alkaline soils are created when sodium is present in organic carbon (HCO_3^- and CO_3^{2-}), whereas saline soils are created when sodium is combined with chloride. As a result, neither soil type is preferred for use in agriculture (Keesari et al., 2016). According to the statistical analysis, the range of values for %Na is 19.58–90.02, with a mean of 47.29 ± 15.41

Table 4.9 Descriptive statistics for measured groundwater quality parameters and irrigation indices

Parameters	Maximum	Minimum	Average	SD
EC (μS/cm)	2002	184	760.53	449.85
SAR	14.63	0.49	2.64	2.46
%Na	90.02	19.99	47.29	15.41
SSP (%)	84.7	18.62	42.4	14.35
RSC (meq/L)	8.22	−8.17	0.88	2.39
MH (%)	86.83	7.35	52.77	18.36
KR	8.49	0.23	1.01	1.37
PI	111.55	46.76	79.2	13.06
EIWQI	192.03	27.42	52.32	27.95

Based on Mahammad et al., 2022

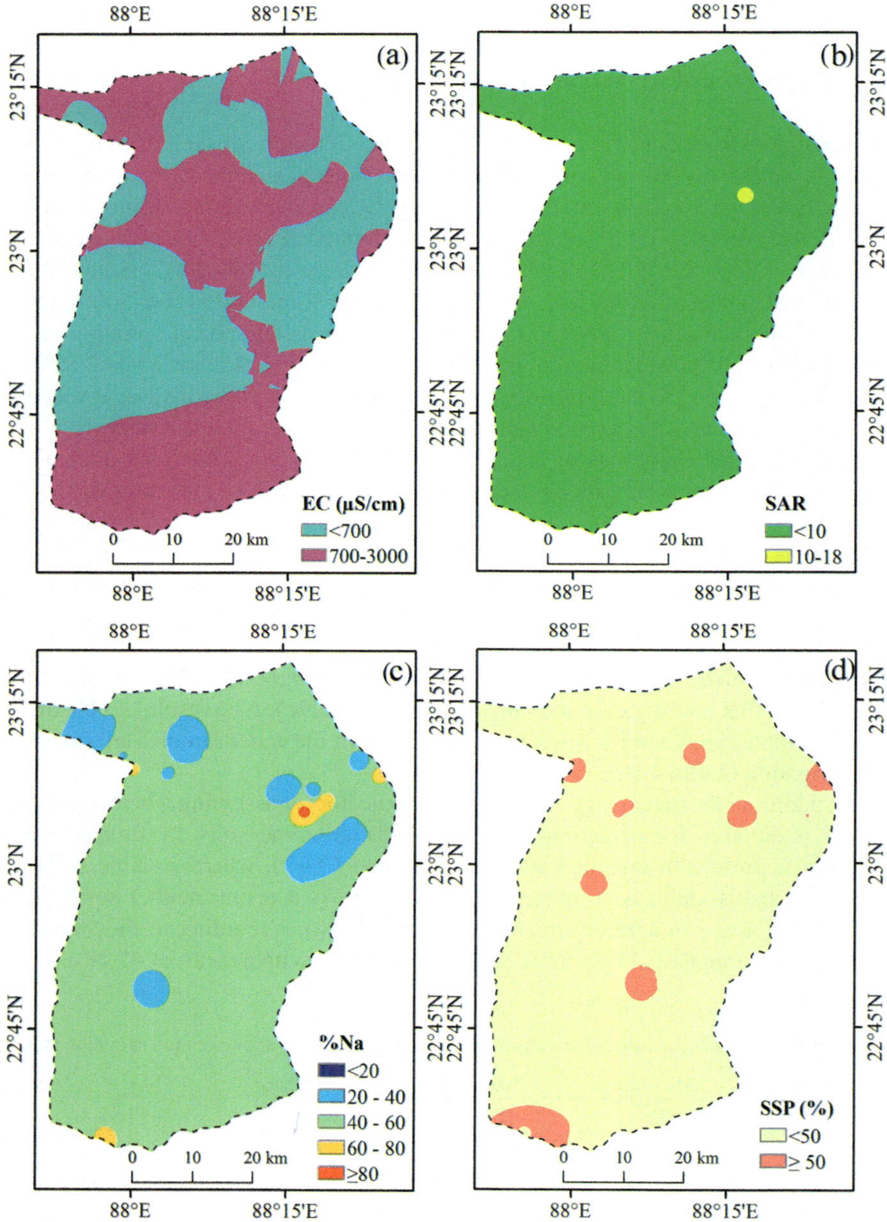

Fig. 4.29 Spatial distribution of groundwater irrigation hazard: (**a**) EC, (**b**) SAR, (**c**) %Na, and (**d**) SSP (Based on Mahammad et al., 2022)

Table 4.10 Rating of the groundwater samples for irrigation based on the EC, SAR, KR, %Na, SSP, RSC, MH, PI indices, and EIWQI

Parameters (unit)	Range	Water category	Number of wells	Percentage of wells
pH	<6.5	Unsuitable	0	0.00
	6.5–8.5	Suitable	32	94.12
	>8.5	Unsuitable	2	5.88
Cl^- (mg/L)	<354	Suitable	33	97.06
	>354	Unsuitable	1	2.94
NO_3^- (mg/L)	<30	Suitable	34	100
	>30	Unsuitable	0	0.00
EC (μS/cm)	<1500	Sensitive	31	97.18
	1500–3000	Semi-tolerant	3	8.82
	>3000	Tolerant	0	0.00
SAR	0–10	Excellent	33	97.06
	10–18	Good	1	2.94
	18–26	Fair	0	0
	>26	Unsuitable	0	0
%Na	0–20	Excellent	1	2.94
	20–40	Good	11	32.35
	40–60	Permissible	17	50
	60–80	Poor	4	11.76
	>80	Unsuitable	1	2.94
SSP (%)	<50	Good	25	73.53
	>50	Unsuitable	9	26.47
RSC (meq/L)	<1.25	Excellent	23	67.65
	1.25–2.5	Good	6	17.65
	>2.5	Unsuitable	5	14.70
MH (%)	<50	Good	13	38.24
	>50	Unsuitable	21	61.76
Kr	<1	Suitable	23	67.65
	>1	Unsuitable	11	32.35
PI	<80	Good	18	52.94
	80–100	Moderate	14	41.18
	100–120	Poor	1	2.94
	>120	Unsuitable	1	2.94
EIWQI	<50	Excellent	18	52.94
	50–75	Good	14	41.18
	75–100	Moderate	1	2.94
	>100	Poor	1	2.94

Based on Mahammad et al., 2022

(Table 4.9). Five groups have been established for the % Na concentration: (1) excellent (0–20), (2) good (20–40), (3) acceptable (40–60), (4) poor (60–80), and (5) unsuitable (>80). The measured groundwater quality is categorized as follows: 2.94% excellent, 32.35% good, 50% permissible, 11.76% poor, and 2.94%

unsuitable for irrigation, according to the regional variance of the %Na (Table 4.10; Fig. 4.29c).

The groundwater samples are categorized into five kinds (Wilcox, 1948) by the Wilcox diagram based on the percentage of EC and Na concentration: (1) excellent to good, (2) good to permissible, (3) permissible to questionable, (4) doubtful to unsuitable, and (5) unsuitable (Fig. 4.30a). According to the study's Wilcox diagram, the best to good class is most common, followed by the good to permitted, permissible to doubtful, and doubtful to inappropriate categories. It is discovered that 58.82% of the groundwater samples belong to the excellent to good class, meaning they are better suited for irrigation purposes. Comparably, 23.53% of the groundwater samples are rated as acceptable to good, meaning they pose less of a risk for usage in irrigation. The percentage of samples with moderate salt hazard that fall into the permitted to dubious category is 17.65%. In addition, 2.94% of the water samples are deemed questionable to inappropriate for use in irrigation. No groundwater sample is classified as unfit for use as irrigation water. As a result, the analysis shows that more than 80% of the groundwater samples in this research region are located in areas that are suitable for irrigation.

An important method for identifying irrigation water based on SAR value and electrical conductivity is the USSL method. Regarding the salinity hazard, all of the groundwater samples in this study are categorized into C1, C2, C3, and C4 and S1, S2, S3, and S4 regarding the sodium hazard. One sample is in the C1 class on the USSL diagram (Fig. 4.30b), which indicates low salinity water that is appropriate for a variety of crops on different types of soils. In contrast, the C2 class is seen in 52.94% of the sample wells. It demonstrates that crops tolerant of a medium salt content can be irrigated with medium salinity water (Balasubramani et al., 2020). Furthermore, 44.12% of the samples express C3, which is regarded as high salinity water. According to this class, plants that can withstand modest salt concentrations can use groundwater (Balasubramani et al., 2020). Additionally, according to Balasubramani et al. (2020), the analysis shows that none of the groundwater samples fit into the high salinity water (C4) group, which necessitates the need of special salinity management measures. Furthermore, 2.94% and 2.94% of the groundwater samples, respectively, have been classified as low (S2) and moderate (S3) salinity hazards, whereas 94.12% of the samples had extremely low salinity hazards (S1). Furthermore, none of the groundwater samples is in the high salinity hazard (S4) zone, similar to the salinity hazard. Thus, it can be concluded that salinity and salt contamination have little effect on groundwater for irrigation. Furthermore, it is discovered that 89.66% of the groundwater samples belong to the S1 class, which comprises low-sodium water. The majority of crops with minimal sodium content can be grown with this water. In groundwater samples, the medium sodium water (S2) category has only been found in 6.9%. When exposed to fine-textured soils with low infiltration capacity, the water samples in this class become susceptible to sodium risks (Balasubramani et al., 2020). Only one sample (GWQ25) fits into the S3 class, which denotes a high concentration of salt in water. None of the samples are deemed to be very high sodium concentration water (S4), which is typically not recommended for agricultural use (Balasubramani et al., 2020).

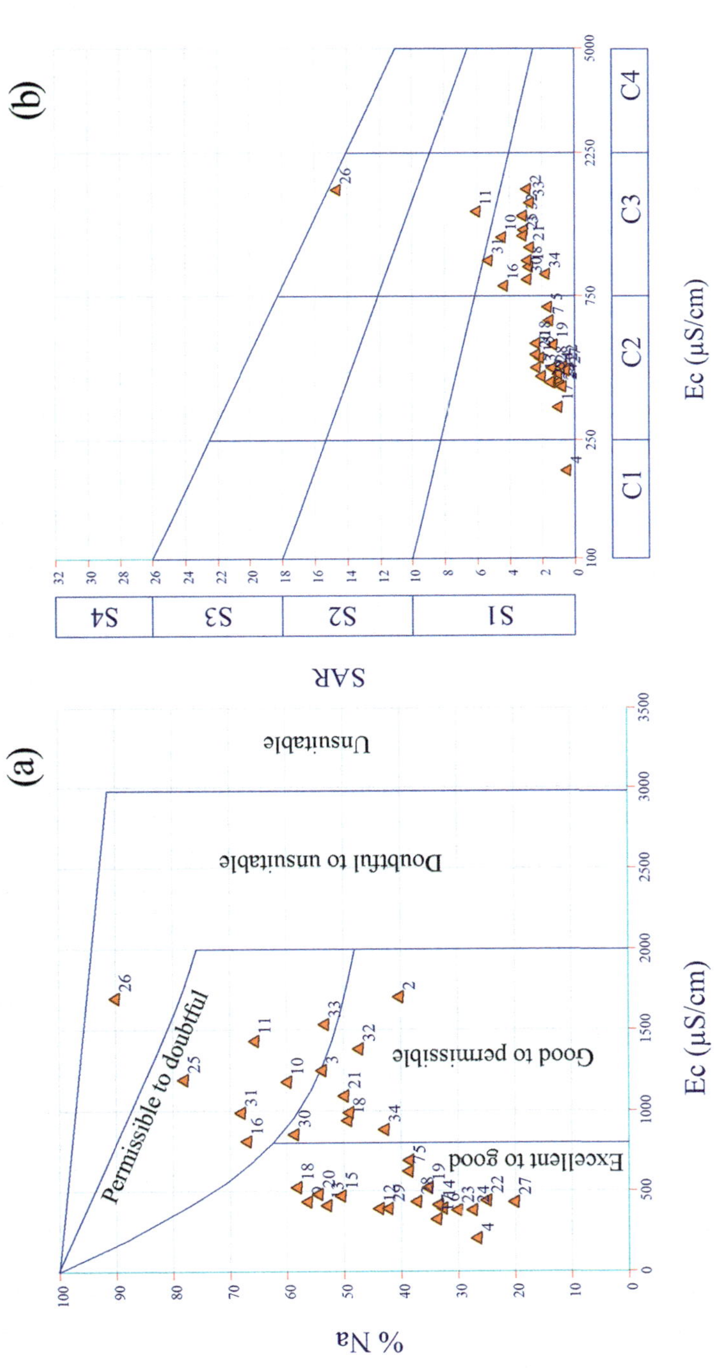

Fig. 4.30 Suitability of groundwater: (**a**) Wilcox diagram and (**b**) salinity and alkalinity hazard (Based on Mahammad et al., 2022)

In order to categorize irrigation water according to soil permeability, the SSP is a crucial indication (Singh et al., 2020). Soil hardening caused by the exchange process of sodium ions with magnesium and calcium in clay particles results in low soil permeability and restricted drainage conditions (Tijani, 1994; Singh et al., 2020). SSP levels above 50 denote unsuitability for irrigation, while values below 50 indicate that groundwater is appropriate for irrigation (Todd & Mays, 2004). SSP average is 42.4 ± 14.35, with a range of 18.62–84.7 (Table 4.9). Two groups have been established for the spatiality of SSP: (1) excellent (<50) and (2) unsuitable (>50). According to the spatial variation of the SSP, 26.47% of the samples are deemed unsuitable for irrigation, whereas 73.53% of the samples are deemed good (Table 4.10, Fig. 4.29d).

According to Mthembu et al. (2021), the RSC is a crucial method for identifying the detrimental impacts of carbonate and bicarbonate on water quality. Water should not be used for irrigation if the RSC content is more than 2.5 (FAO, 1985). Consequently, three groups have been established based on RSC concentration: (1) excellent (<1.25), (2) good (1.25–2.5), and (3) unsuitable (>1.25). In the DFD, the RSC concentration ranges from −8.17 to 8.22, with an average of 0.88 ± 2.89 (Table 4.9). Approximately 17.64% of the sample water has good irrigation adaptability and 67.65% of the sampled groundwater has exceptional appropriateness (Table 4.10). Conversely, 14.71% of the groundwater analysed is not suitable for irrigation (Fig. 4.31a).

A crop's yield could be negatively impacted by magnesium in the groundwater (Gowd, 2005). In this study region, the MH values had a maximum of 86.83, a minimum of 7.35, an average of 52.77, and a standard deviation of 18.36%, as indicated (Table 4.9). A threshold value of 50 has been proposed for the MH in irrigation (Narasaiah & Rao, 2021a, b). It is therefore inappropriate for irrigation if the MH content is more than 50. There are two groups for MH concentration: (1) excellent (<50) and (2) inappropriate (>50) (Table 4.10). According to the spatiality of the MH, 38.24% of the samples are deemed good, whereas 61.76% are deemed unsuitable for irrigation (Fig. 4.31b).

Plant growth is retarded by the high content of salt in irrigation water (Kelley et al., 1940). Water becomes unfit for irrigation when KR rises above 50. Table 4.10 revealed that the KR value ranged from 0.23 to 8.49, with an average of 1.01 ± 1.37. Two types are observed in relation to KR: appropriate (<1) and unsuitable (>1). A little over 67.65% of the samples are in the appropriate class, and 32.35% are not appropriate for irrigation (Table 4.10, Fig. 4.31c).

Table 4.9 revealed that the PI ranged from 46.76 to 111.55, with an average of 79.2 ± 13.06. The range of PI is classified as good (<80), moderate (80–100), poor (100–120), and unsuitable (>120) by Doneen (1964). For irrigation reasons, 52.94% of the samples are deemed good, 41.18% moderate, 2.94% poor, according to the geographical variation of the PI (Fig. 4.31d). Regarding the PI, 2.94% of the sample wells in this instance was judged to be unsuitable (Table 4.10).

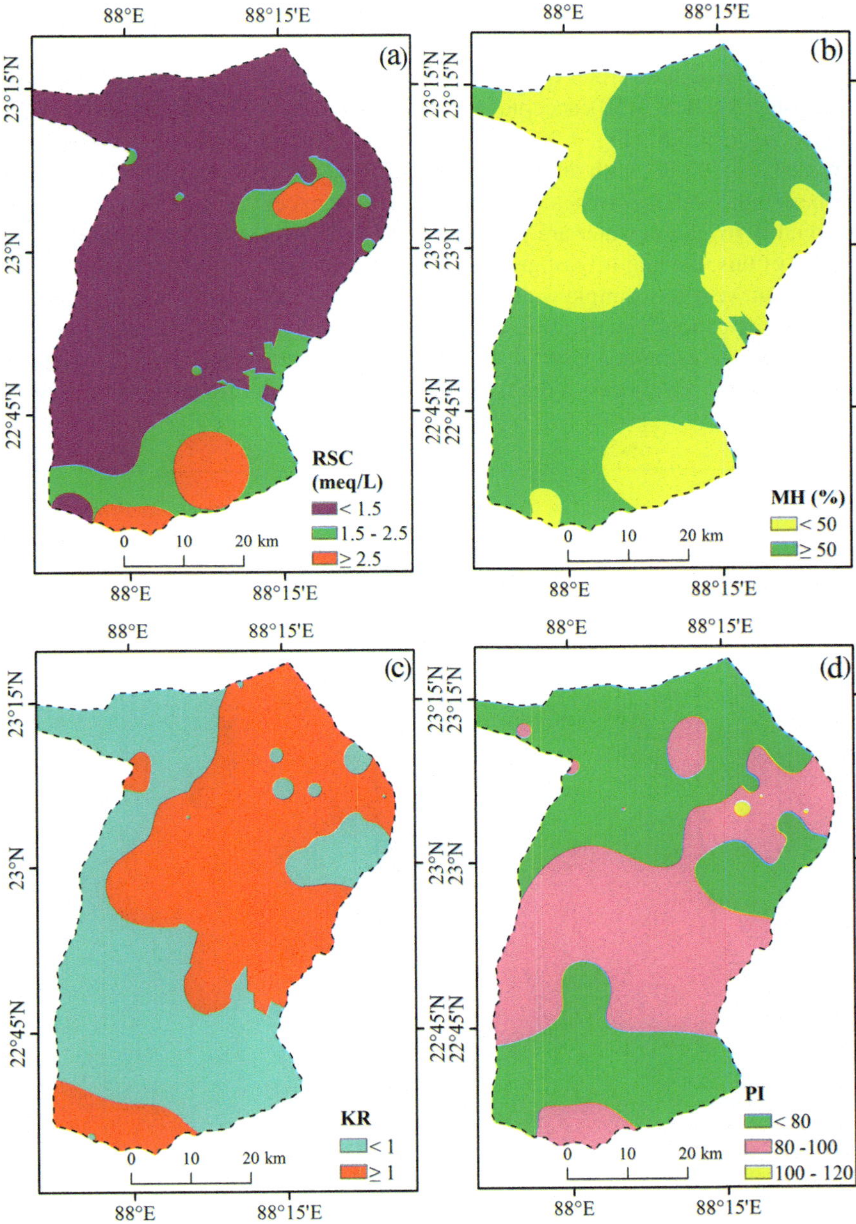

Fig. 4.31 Spatial distribution groundwater irrigation hazard: (**a**) RSC, (**b**) MH, (**c**) KR, and (**d**) PI (Based on Mahammad et al., 2022)

4.4.2.2 EIWQI

Based on the irrigation water quality index value, the EIWQI range is divided into five classes: excellent (<50), acceptable (50–75), moderate (75–100), bad (100–125), and very poor (>125) (Fig. 4.32). Table 4.9 revealed that the EIWQI value ranged from 27.42 to 192.03, with an average of 52.32 ± 27.95. The EIWQI results show that the sample wells' corresponding percentages of excellent, good, moderate, and poor quality irrigation water are 52.94%, 41.18%, 2.94%, and 2.94% (Table 4.10). Because of this, the majority of groundwater samples fall into the excellent to good category, but very few sample wells are deemed to be of extremely poor to middling quality for irrigation. The majority of the DFD exhibits an excellent water category, according to the estimated EIWQI. In addition, the northern and central regions of the research area have access to high-quality water. The intermediate class of EIWQI

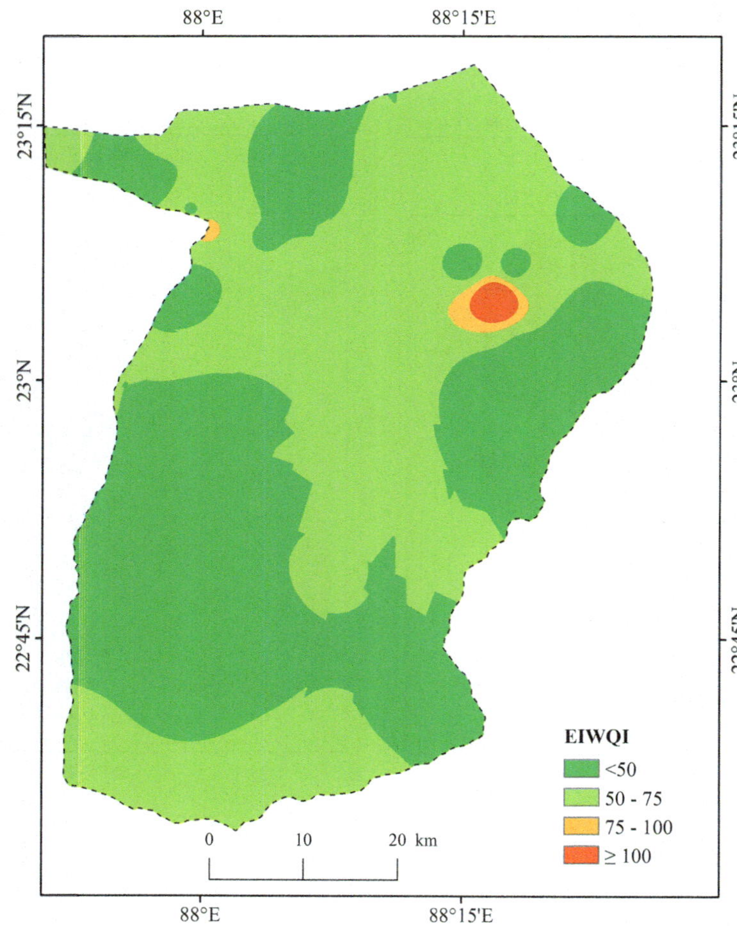

Fig. 4.32 Spatial distribution of Irrigation Water Quality Index

is seen in a circular shape around well 11. Furthermore, the sample well W26, which is situated in the eastern portion of the DFD, has low-quality water. In certain instances, the spatial distribution of the parameters used to derive the EIWQI and the EIWQI is comparable.

4.5 Conclusions

Throughout this chapter, the spatiotemporal dynamics of the surface water resources along with its quality assessment are examined. In the PRM period, the expansion of the waterbody class attained 0.17% growth from 1989 to 2020. Moreover, in the POM period, the area covered by waterbody is gradually decreasing—0.43% during the study period. The seasonal distribution of rainfall is closely associated with the distribution of the waterbodies in the study area. The transitional intensity in the dynamics of the waterbody for both the PRM and POM seasons, agricultural fallow and, agricultural land and vegetation cover has been considered as the dominant contributor. The alteration of agricultural activity and seasonal growth of vegetation cover on the waterbody has been observed. Apart from the qualitative assessment, the surface water quality assessment indicates that there is a spatial variation. The surface water quality for drinking purpose (WQI) ranges from 50 to 100, which indicates good quality. The spatial variation shows that the lower value of WQI, which indicates better water quality, is confined to the upper part of the study area. Regarding irrigation water, quality areas show that overall irrigation water quality is better in the upper part of the study area compared to its lower part. The result illustrated that physical and anthropogenic activity is responsible for the deterioration of surface water quality.

The temporal variation of the study implies that except for a few locations, most of the wells have a decreasing trend in the groundwater depth. The spatial distribution of GWL revealed that maximum changes of GWL occurred in the POM *kharif* season followed by PRM, MON, and POM *rabi* seasons. The predicted GWL of 2025–2026 shows declining trend in groundwater levels, especially in the western part of the study area around Barddhaman town. This GWL fall leads to dwindling agricultural economy.

Along with the depletion of the groundwater level, this inquiry attempted to evaluate the GWQ for drinking and irrigation purposes. For this purpose, the research used physicochemical parameters, major anions, cations, and heavy metals. The WQI of groundwater shows that 50% of the sample is excellent, while 11.76% of the samples are very poor. The spatial distribution of the EIWQI reveals that most of the samples of this study area are of excellent and very good quality and distributed in the eastern, western, and southern fringe areas. The spatial pattern of the EIWQI appears similar to the alignment with the spatial variation of WQI. A high concentration of EC, Cl^-, and Na^+ resulted from the increasing salinity due to high evaporation and anthropogenic activities.

The depiction of surface water availability and fall of groundwater level is really crucial to know the water surplus and deficit areas of the DFD; therefore, the next chapter is concerned with the assessment of the flood and drought scenario of the DFD.

References

Abbas, Z., Zhu, Z., & Zhao, Y. (2022). Spatiotemporal analysis of landscape pattern and structure in the Greater Bay Area, China. *Earth Science Informatics, 15*(3), 1977–1992.

Acharya, T. D., Subedi, A., & Lee, D. H. (2018). Evaluation of water indices for surface water extraction in a Landsat 8 scene of Nepal. *Sensors, 18*(8), 2580.

Adimalla, N., Li, P., & Venkatayogi, S. (2018). Hydrogeochemical evaluation of groundwater quality for drinking and irrigation purposes and integrated interpretation with water quality index studies. *Environmental Processes, 5*, 363–383.

Adimalla, N., Qian, H., & Li, P. (2020). Entropy water quality index and probabilistic health risk assessment from geochemistry of groundwaters in hard rock terrain of Nanganur County, South India. *Geochemistry, 80*(4), 125544.

Ahada, C. P., & Suthar, S. (2018). Assessing groundwater hydrochemistry of Malwa Punjab, India. *Arabian Journal of Geosciences, 11*, 1–15.

Arkoprovo, B., Adarsa, J., & Prakash, S. S. (2012). Delineation of groundwater potential zones using satellite remote sensing and geographic information system techniques: A case study from Ganjam district, Orissa, India.

Aydin, H., Ustaoğlu, F., Tepe, Y., & Soylu, E. N. (2021). Assessment of water quality of streams in Northeast Turkey by water quality index and multiple statistical methods. *Environmental Forensics, 22*(1–2), 270–287.

Balasubramani, K., Murthy, K. R., Gomathi, M., & Kumaraswamy, K. (2020). Integrated assessment of groundwater resources in a semi-arid watershed of South India: implications for irrigated agriculture. *GeoJournal, 85*(6), 1701–1723. https://doi.org/10.1007/s10708-019-10050-0

Bijeesh, T. V., & Narasimhamurthy, K. N. (2020). Surface water detection and delineation using remote sensing images: A review of methods and algorithms. *Sustainable Water Resources Management, 6*, 1–23.

Binh, N. A., et al. (2021). Thirty-year dynamics of LULC at the Dong Thap Muoi area, southern Vietnam, using Google earth engine. *ISPRS International Journal of Geo-Information, 10*(4), 226.

Bozdağ, A. (2015). Combining AHP with GIS for assessment of irrigation water quality in Çumra irrigation district (Konya), Central Anatolia, Turkey. *Environmental Earth Sciences, 73*(12), 8217–8236.

Breiman, L. (2001). Random forests. *Machine Learning, 45*, 5–32.

Chang, L. Y., & Chien, J. T. (2013). Analysis of driver injury severity in truck-involved accidents using a non-parametric classification tree model. *Safety Science, 51*(1), 17–22.

Chen, X., Wang, M., & Zhang, H. (2011). The use of classification trees for bioinformatics. *Wiley Interdisciplinary Reviews: Data Mining and Knowledge Discovery, 1*(1), 55–63.

Das, S., & Sarkar, R. (2019). Predicting the land use and land cover change using Markov model: A catchment level analysis of the Bhagirathi-Hugli River. *Spatial Information Research, 27*, 439–452.

Doneen, L. D. (1964). *Water quality for agriculture* (p. 48). Department of Irrigation, University of California.

Dutta, D., et al. (2020). Linking LULC change with urban heat islands over 25 years: A case study of the urban-industrial city Durgapur, eastern India. *Journal of Spatial Science, 65*(3), 501–518.

FAO. (1985). *Water quality for agriculture*. Food and Agriculture Organization of the United Nations.

Feizizadeh, B., Omarzadeh, D., Kazemi Garajeh, M., Lakes, T., & Blaschke, T. (2021). Machine learning data-driven approaches for land use/cover mapping and trend analysis using Google Earth Engine. *Journal of Environmental Planning and Management, 66*(3), 665–697.

Feyisa, G. L., Meilby, H., Fensholt, R., & Proud, S. R. (2014). Automated water extraction index: A new technique for surface water mapping using Landsat imagery. *Remote Sensing of Environment, 140*, 23–35.

Ghosh, S., Dasgupta, A., & Swetapadma, A. (2019). A study on support vector machine based linear and non-linear pattern classification. In *2019 international conference on intelligent sustainable systems (ICISS)* (pp. 24–28). IEEE.

Gowd, S. S. (2005). Assessment of groundwater quality for drinking and irrigation purposes: A case study of Peddavanka watershed, Anantapur District, Andhra Pradesh, India. *Environmental Geology, 48*(6), 702–712. https://doi.org/10.1007/s00254-005-0009-z

Hamlat, A., & Guidoum, A. (2018). Assessment of groundwater quality in a semiarid region of northwestern Algeria using water quality index (WQI). *Applied Water Science, 8*(8), 1–13.

Keesari, T., Ramakumar, K. L., Chidambaram, S., Pethperumal, S., & Thilagavathi, R. (2016). Understanding the hydrochemical behavior of groundwater and its suitability for drinking and agricultural purposes in Pondicherry area, South India–a step towards sustainable development. *Groundwater for Sustainable Development, 2*, 143–153. https://doi.org/10.1016/j.gsd.2016.08.001

Kelley, W. P., Brown, S. M., & Liebig, G. F., Jr. (1940). Chemical effects of saline irrigation water on soils. *Soil Science, 49*(2), 95–108.

Kombo, O. H., Kumaran, S., Sheikh, Y. H., Bovim, A., & Jayavel, K. (2020). Long-term groundwater level prediction model based on hybrid KNN-RF technique. *Hydrology, 7*, 59. https://doi.org/10.3390/hydrology7030059

Krishnamurthy, J., Mani, A., Jayaraman, V., & Manivel, M. (2000). Groundwater resources development in hard rock terrain-an approach using remote sensing and GIS techniques. *International Journal of Applied Earth Observation and Geoinformation, 2*(3–4), 204–215.

Mahammad, S., & Islam, A. (2021). Evaluating the groundwater quality of Damodar Fan Delta (India) using fuzzy-AHP MCDM technique. *Applied Water Science, 11*(7), 1–17. https://doi.org/10.1007/s13201-021-01408-2

Mahammad, S., Islam, A., & Shit, P. K. (2022). Geospatial assessment of groundwater quality using entropy-based irrigation water quality index and heavy metal pollution indices. *Environmental Science and Pollution Research, 30*, 116498. https://doi.org/10.1007/s11356-022-20665-5

Mahammad, S., Islam, A., Shit, P. K., Towfiqul Islam, A. R. M., & Alam, E. (2023). Groundwater level dynamics in a subtropical fan delta region and its future prediction using machine learning tools: Sustainable groundwater restoration. *Journal of Hydrology: Regional Studies, 47*, 101385. https://doi.org/10.1016/j.ejrh.2023.101385

Manzoor, S. A., Griffiths, G. H., Robinson, E., Shoyama, K., & Lukac, M. (2022). Linking pattern to process: Intensity analysis of land-change dynamics in Ghana as correlated to past socioeconomic and policy contexts. *Land, 11*(7), 1070.

Margat, J., & Van der Gun, J. (2013). *Groundwater around the world: A geographic synopsis*. CRC Press.

Minh, H. V. T., Avtar, R., Kumar, P., Tran, D. Q., Ty, T. V., Behera, H. C., & Kurasaki, M. (2019). Groundwater quality assessment using fuzzy-AHP in an Giang Province of Vietnam. *Geosciences, 9*(8), 330.

Mthembu, P. P., Elumalai, V., Senthilkumar, M., & Wu, J. (2021). Investigation of geochemical characterization and groundwater quality with special emphasis on health risk assessment in alluvial aquifers, South Africa. *International journal of Environmental Science and Technology, 1-20*, 3711. https://doi.org/10.1007/s13762-021-03129-0

Narasaiah, V., & Rao, B. V. (2021a). Groundwater quality of a hard rock aquifer in the Subledu Basin of Khammam district, India. *Applied Water Science, 11*(6), 1–21. https://doi.org/10.1007/s13201-021-01424-2

Narasaiah, V., & Rao, B. V. (2021b). Groundwater quality of an hard rock aquifer in the Subledu Basin of Khammam district, India. *Applied Water Science, 11*(6), 89.

Panigrahi, A. K., & Pattnaik, S. (2020). Water pollution effects of river Saraswati and Kunti on livelihood: A case study of fishermen communities in Tribeni and Kuntighat, Hooghly, West Bengal. *Journal of Fisheries, 8*(3), 928–934.

Pathak, S., Mishra, I., & Swetapadma, A. (2018, November). An assessment of decision tree based classification and regression algorithms. In *2018 3rd international conference on inventive computation technologies (ICICT)* (pp. 92–95). IEEE.

Rahaman, S., Kumar, P., Chen, R., Meadows, M. E., & Singh, R. B. (2020). Remote sensing assessment of the impact of land use and land cover change on the environment of Barddhaman district, West Bengal, India. *Frontiers in Environmental Science, 8*, 127.

Saputra, M. H., & Lee, H. S. (2019). Prediction of land use and land cover changes for North Sumatra, Indonesia, using an artificial-neural-network-based cellular automaton. *Sustainability, 11*(11), 3024.

Seifi, A., Ehteram, M., Singh, V.P., & Mosavi, A. (2020). Modeling and uncertainty analysis of groundwater level using six evolutionary optimization algorithms hybridized with ANFIS, SVM, and ANN. *Sustainability 12,* 4023. https://doi.org/10.3390/su12104023.

Singh, K. K., Tewari, G., & Kumar, S. (2020). Evaluation of groundwater quality for suitability of irrigation purposes: A case study in the Udham Singh Nagar, Uttarakhand. *Journal of Chemistry, 2020*. https://doi.org/10.1155/2020/6924026

Subramani, T., Elango, L., & Damodarasamy, S. R. (2005). Groundwater quality and its suitability for drinking and agricultural use in Chithar River Basin, Tamil Nadu, India. *Environmental Geology, 47*(8), 1099–1110. https://doi.org/10.1007/s00254-005-1243-0

Tijani, M. N. (1994). Hydrogeochemical assessment of groundwater in Moro area, Kwara State, Nigeria. *Environmental Geology, 24*, 194–202.

Todd, D. K., & Mays, L. W. (2004). *Groundwater hydrology*. John Wiley & Sons.

Tyagi, S., Sharma, B., Singh, P., & Dobhal, R. (2013). Water quality assessment in terms of water quality index. *American Journal of Water Resources, 1*(3), 34–38. https://doi.org/10.12691/ajwr-1-3-3

USSL (1954). Diagnosis and improvement of saline and alkali soils, United States Department of Agriculture, Agricultural handbook no.60, Washington, D.C, 147p.

Wagh, V. et al. (2017). Groundwater quality evaluation by physicochemical characterization and water quality index for Nanded Tehsil, Maharashtra, India. In *Proceedings of the 7th international conference on biology, environment and chemistry, IPCBEE* (Vol. 98).

Wang, B., Fang, B., Wang, Y., Liu, H., & Liu, Y. (2016a). Power system transient stability assessment based on big data and the core vector machine. *IEEE Transactions on Smart Grid, 7*(5), 2561–2570.

Wang, L. A., Zhou, X., Zhu, X., Dong, Z., & Guo, W. (2016b). Estimation of biomass in wheat using random forest regression algorithm and remote sensing data. *The Crop Journal, 4*(3), 212–219. https://doi.org/10.1016/j.cj.2016.01.008

Warmuth, M. K., Liao, J., Rätsch, G., Mathieson, M., Putta, S., & Lemmen, C. (2003). Active learning with support vector machines in the drug discovery process. *Journal of Chemical Information and Computer Sciences, 43*(2), 667–673.

Wilcox, L. (1948). *The quality of water for irrigation use (Technical Bulletin 1962)*. United State Department of Agriculture.

Chapter 5
Delineation of Water Surplus and Deficit Areas

5.1 Backdrop

Water resource management is vital especially in the era of the Anthropocene under the threat of climate change because hydrological extremes that are floods and droughts have exacerbated with time. Thus, the management of water resources embodies the idea of managing droughts and floods in an integrated manner.

Water resource management in a water-stressed area like the Damodar fan delta (DFD) involves two aspects: (1) increasing the supply of water and (2) decreasing the demand for water for various purposes. There are many classical studies (Howe, 1976; Hertel & Liu, 2019; Boretti & Rosa, 2019; Kumar et al., 2008) that emphasized increasing water supply for agricultural development and economic prosperity. However, increasing water supply in the absolute term is not possible (Falkenmark, 1989), especially in recent times. Thus, managing demand with the quantity of water is more practical. Naturally, many scientists focus on demand-side management of water resources across the world (Mirdashtvan et al., 2021; Allegretti et al., 2022; Usman et al., 2022). In the context of the DFD, these two types of management techniques have some scope and limitations. As this study area is plagued with some inherent problems of riverine floods and recurring droughts, the management of floodwater diversion to the water-stressed areas tends to elevate the quantum of water resources in the area on the one hand and reduce flood on the other. Moreover, there are many palaeochannels of the DFD that have some potential for groundwater recharge through reducing the surface run-off. Thus, palaeochannels may play instrumental role in shaping the water resource management of the DFD. Thus, this chapter is devoted to knowing the nature of water surplus areas due to monsoonal floods and water deficit areas due to drought. Besides the potentiality of the palaeochannels in groundwater recharge, the direction and dimensions of the recharge area are also needed.

S. Mahammad, A. Islam, *Palaeochannels and Water Resource Management*,
https://doi.org/10.1007/978-3-031-92021-9_5

5.2 Assessment of Water Surplus Area Due to Flood

Damodar River is known as the 'Sorrow of Bengal' due to its history of causing devastating floods (Rudra et al., 2018; Ghosh & Mistri, 2012). With varying magnitudes and recurrent intervals of 8–10 years, the Damodar River has been causing devastation since the first recorded flood occurrence in 1730 up until 2007 (Bhattacharyya, 2011). This study area experienced major flood events in 1978, 1984, 1987, 1995, 1996, 1998, 1999, 2000, 2006, 2007, and 2009 and moderate floods happened in 1980, 1985, 1986, 1990, 1991, 1993, 1994, 2011, 2013, and 2015 (Das et al., 2017). It is also found that flood events occur frequently in the lower reaches of Damodar River comprising the lower part of the study area. This chapter focuses on extracting flood inundation in the study area using synthetic aperture radar (SAR) data from 2015 to 2019.

5.2.1 *Datasets*

In this study, flood inundation area has been delineated using the Sentinel 1A (S1A) radio detection and ranging (RADAR) data provided by European Space Agency (ESA). The flood events occur in the study area during the monsoon period due to excessive rainfall. Therefore, S1A data has been implemented due to their ability to acquire information in all-weather conditions. The S1A data were used in interferometric wide (IW) and ground range detection format (GRD), which is the multilook product and projected from the slant range to the ground range (Moharrami et al., 2021). The S1A is characterized by dual polarization, including VV and VH polarizations in which VV polarization has a better contrast between water and non-water pixel (Melkamu et al., 2022). Therefore, the extraction of flood inundation area has been done using VV polarization due to higher accuracy (Tripathi et al., 2020). For extraction of the flood inundation area, the S1A VV images have been used from 25.07.2015 to 18.08.2015, 25.08.2016 to 05.09.2016, 24.07.2017 to 05.08.2017, and 26.09.2019 to 06.10.2019. In the study area, 2018 is considered a non-flood year. Therefore, the S1A datasets of 25.07.2018 to 18.08.2018, 25.08.2018 to 05.09.2018, 24.07.2018 to 05.08.2018, and 26.09.2018 to 06.10.2018 have been used for the delineation of normal waterbodies.

5.2.2 *Methodology*

The flood inundation area was delineated using the S1A GRD IW datasets in the Google Earth Engine (GEE) platform. The preprocessing of the S1A SAR data includes orbital noise correction, thermal noise correction, radiometric correction, and terrain correction to eliminate the noises regarding orbital, sensor, quality, and projection systems (Tiwari et al., 2020). The Lee spackle filter with a kernel window size of 7 × 7 has been applied to remove the spackle noise (Mahammad & Islam, 2021). After the preprocessing, the flood inundation area was extracted by threshold value observed from the histogram of the datasets (Agnihotri et al., 2019). Moreover, the normal waterbody has been extracted from the S1A SAR data of the same date range.

5.2.3 Pattern of Flood and Related Processes

In 2015, the flood event occurred from 27 July to 8 August. During this period, extraction of flood inundation area has been conducted using SAR datasets. The results illustrated that the dark pixels demonstrated the water logging condition as the waterbody reflects small backscatter. The result depicts that flood inundation area was mainly found in the lower part of the study area particularly in Singur, Polba-Dadpur, Pursurah, Khanakul I, Khanakul II, Jangipara, Udaynarayanpur, Amta I, Amta II, and Jagatballavpur community development (CD) blocks (Fig. 5.1). The spatial distribution of the flood events reveals the areal expansion of 76.49 km^2 in the study area. The flood event occurred due to heavy rainfall from 25 July to 2 August 2015 causing high run-off in the catchment area of the Damodar River (Irrigation and Waterways Directorate, 2015). As a result, the rivers flowing through the study area carried heavy discharge and flowed above the extreme danger level (EDL) during this period. At Muchighata gauge station, the Hurhura Nadi attained 81.5 m peak flood discharge, which is 1.39 m above the EDL (6.76) (Irrigation and Waterways Directorate, 2015). Moreover, the maximum water level of the lower Damodar River was measured as 14.11 m at the Chanpadanga gauge station, which crossed the EDL (13.5 m) of the gauge station. Besides, the Amta canal flowed with a high peak discharge of 6.75 m at Amta Gauge station above the EDL (6.24 m) (Irrigation and Waterways Directorate, 2015).

The flood inundation area of 2016 has been delineated using the S1A SAR images (Fig. 5.2). Besides, the waterbody of the non-flood period has been shown. The result depicts that the flood condition occurred from 22 August to 5 September 2016. During this period, the flood inundation was confined to a few Mouza of the Khanakul I and II and Udaynarayanpur CD blocks. The devastating rainfall in the catchment area of the Damodar River increased surface run-off. Additionally, released water from the Damodar Valley Corporation (DVC) reservoir is responsible for the flooding. On 22 August 2016, 86,955 cusec water was released from the Durgapur barrage (Irrigation and Waterways Directorate, 2016), which increased the water level of the lower Damodar River. After that, on 23 August, the flood peak discharge crossed the EDL and attained a water level of 13.5 m at Chanpadanga gauge station. Moreover, the flow of Hurhura Nadi crossed the EDL of the Muchighata gauge station (Irrigation and Waterways Directorate, 2016).

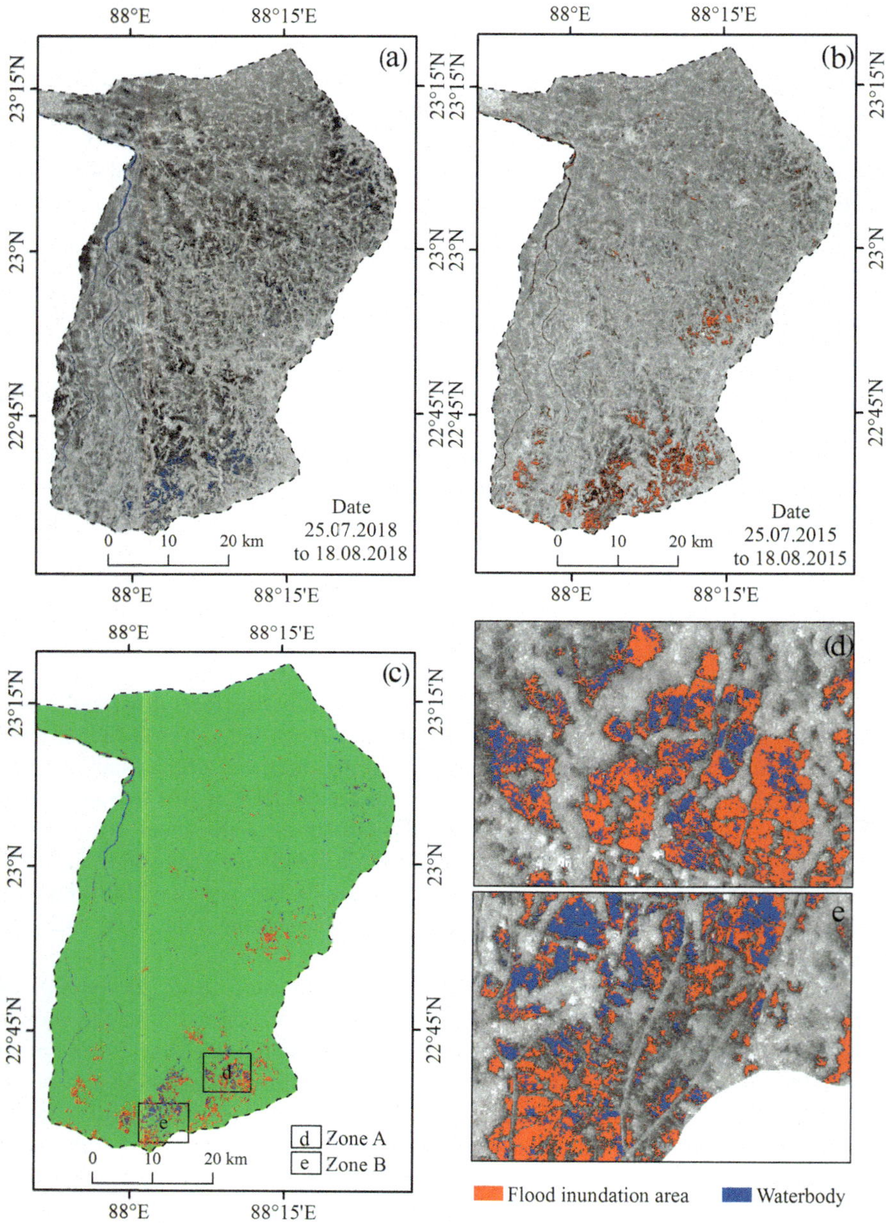

Fig. 5.1 Flood condition of 2015: (**a**) non-flood year, (**b**) flood year, (**c**) distribution of flood inundation area and waterbody, and (**d**) and (**e**) zone of flood inundation area

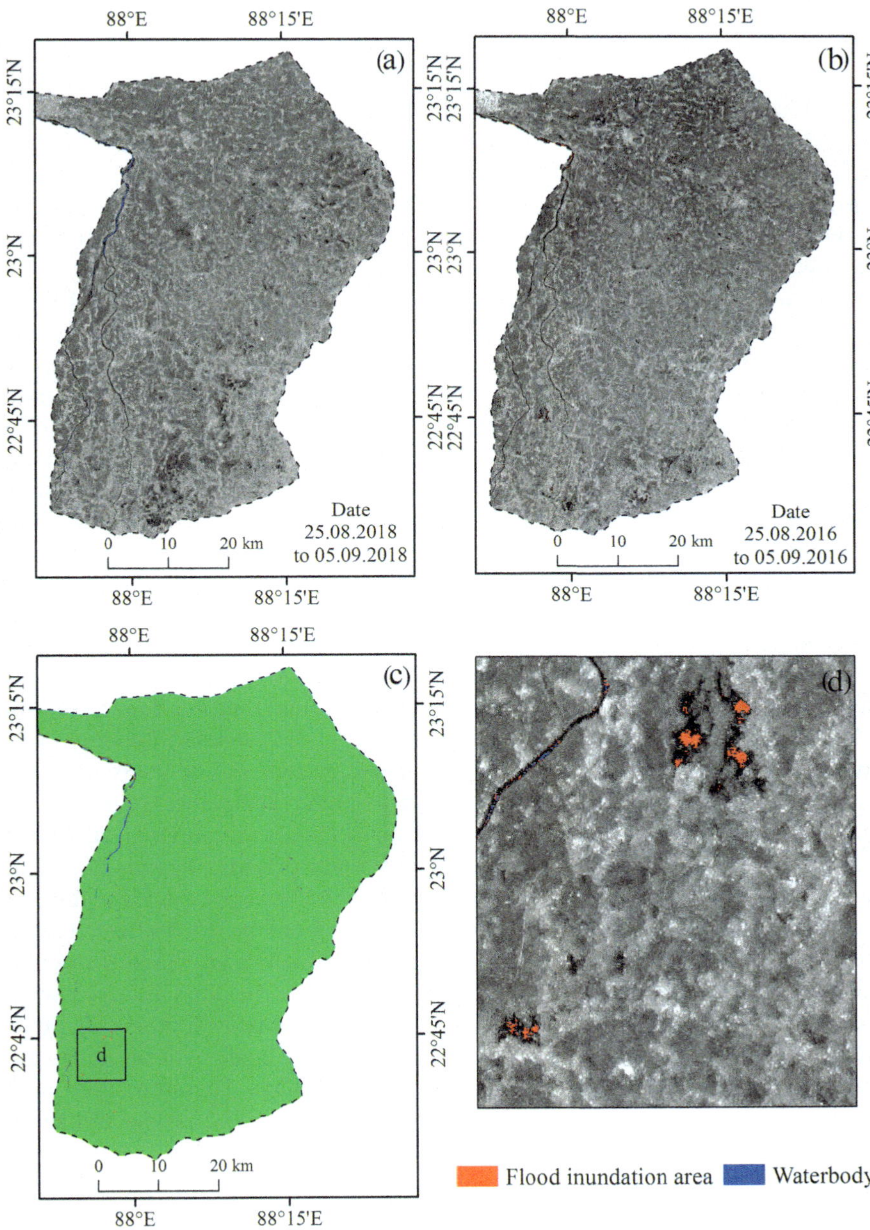

Fig. 5.2 Flood condition of 2016: (**a**) non-flood year, (**b**) flood year, (**c**) distribution of flood inundation area and waterbody, and (**d**) zone of flood inundation area

The water logging area of the 2017 flood has been traced in the present section (Fig. 5.3). It has appeared that the inundation area exhibits 19.84 km^2 in the lower part of the study area, including Khanakul I, Khanakul II, Jangipara, Udaynarayanpur, Amta I, and Amta II CD blocks. The flood event of 2017 occurred due to heavy rainfall precipitating from 21 July to 30 July 2017. Additionally, water released from the DVC reservoir stimulated the flood event in the study area. As a result, the maximum water level attained 23.2, 14.8, 14.5, and 8.68 m at the gauge station of Jamalpur, Chanpadanga, Harinkhola, and Muchighata, respectively (Irrigation and Waterways Directorate, 2017).

The flood event of 2019 happened from 25 September to 5 October 2019. The areal extension of 11.47 km^2 has been traced as the flood inundation area (Fig. 5.4). The result depicts that the flood-affected area is confined mainly to the Khanakul I and Udaynarayanpur CD blocks located in the lower part of the study area. Similar to the previous year, the flood event of 2019 occurred due to heavy rainfall. During the last week of September, the catchment area of the Damodar River received heavy rainfall and generated high run-off. On 25 September, Chanpadanga and Domjur gauge stations received 72.25 mm and 76 mm rainfall (Irrigation and Waterways Directorate, 2019). After that, on 1 October, Chanpadanga and Muchighata gauge stations got a flood peak of 14.25 and 6.88 m exceeding the EDL and inundated the large area (Irrigation and Waterways Directorate, 2019).

This analysis reveals that flood event occurs frequently and it is more devastating in the lower part of the study area. It is also found that the flood condition takes place in the monsoon period (July–October). The causes of flooding in the study region include natural and anthropogenic activities. The study area gets extensive rainfall (80%) in the monsoon period (June–September) (Ghosh et al., 2022). However, the erratic nature of monsoons causes variations in rainfall in this region. Therefore, intensive rainfall in sudden periods happens with high run-off and inundated land areas. Moreover, the tidal lockage of the Hooghly and Rupnarayan Rivers persists to drain the floodwater from the flood-affected areas. Besides, the study area is dominated by floodplain floodwater, which is stored in the low land in stagnant condition (Das et al., 2017). Besides, released water from the DVC reservoir and dams in peak periods stimulates flood events. Apart from that, transforming waterbodies into agricultural land, built-up areas, and other infrastructure acts as a barrier to floodwater drainage. Therefore, this region needs water resource management through conservation and proper utilization of water.

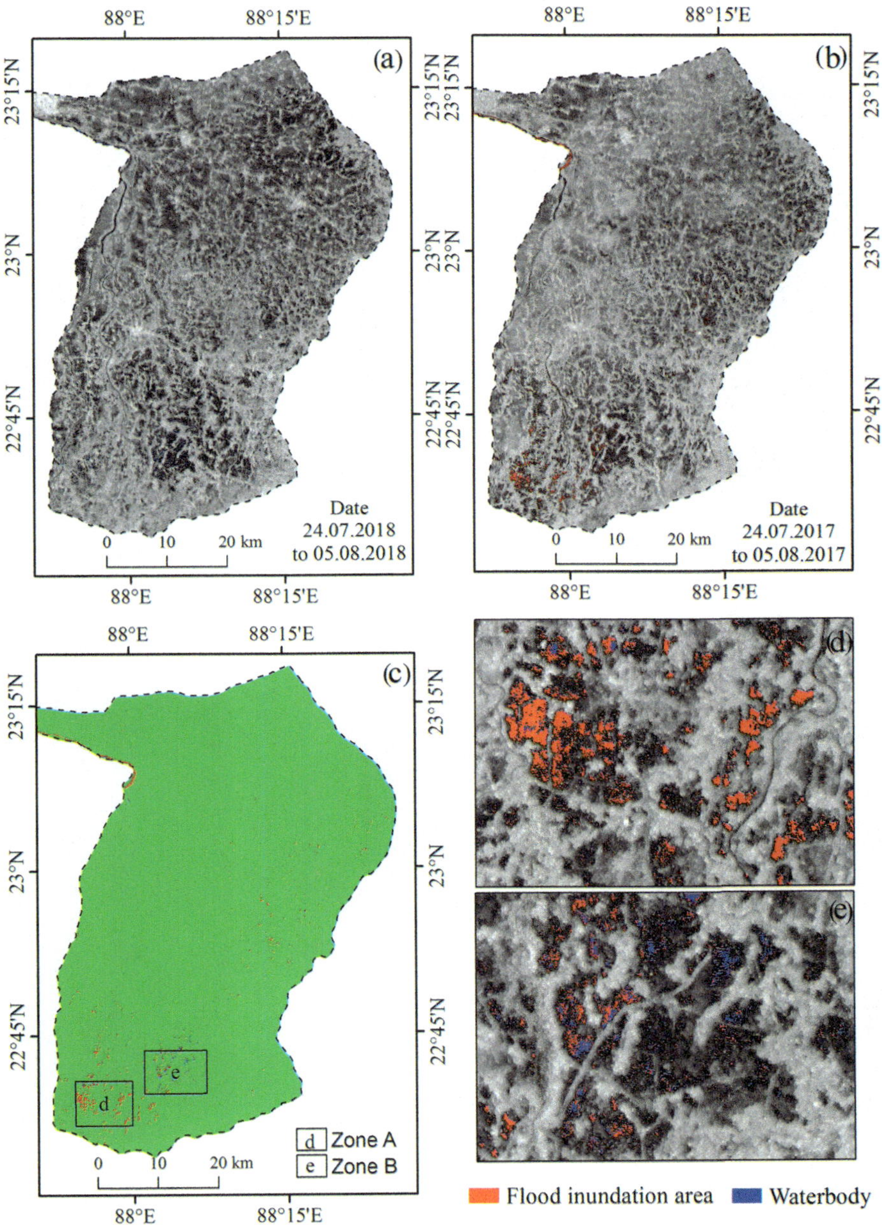

Fig. 5.3 Flood condition of 2017: (**a**) non-flood year, (**b**) flood year, (**c**) distribution of flood inundation area and waterbody, and (**d**) and (**e**) zone of flood inundation area

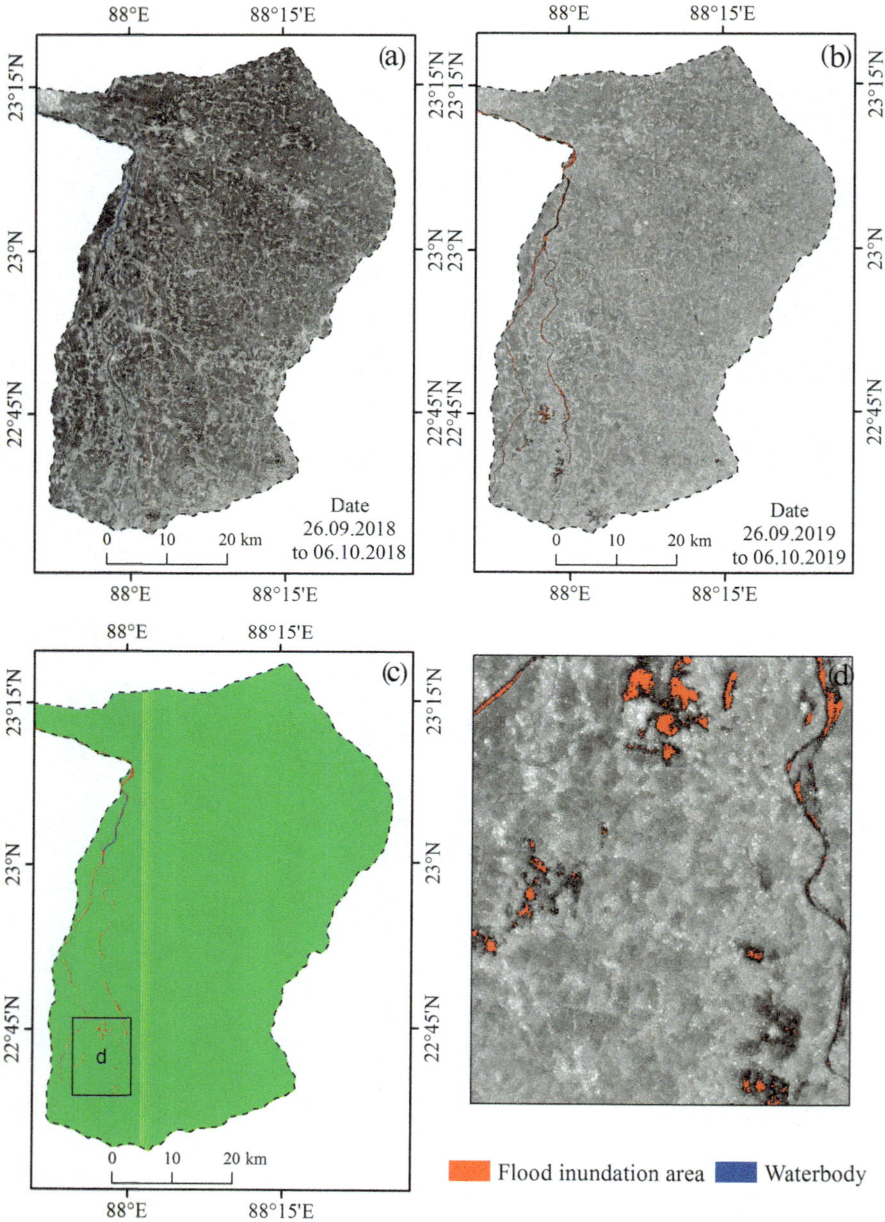

Fig. 5.4 Flood condition of 2019: (**a**) non-flood year, (**b**) flood year, (**c**) distribution of flood inundation area and waterbody, and (**d**) zone of flood inundation area

5.3 Assessment of Water Deficit Area Due to Drought

5.3.1 Datasets

In this study, the drought condition of pre-monsoon, monsoon, post-monsoon *rabi,* and post-monsoon *kharif* seasons of 2010 and 2020 has been assessed. In this research, drought conditions have been analysed based on a meteorological perspective using standardized precipitation perspective (SPI), vegetation health perspective using vegetation condition index (VCI), soil moisture perspective using moisture adequacy index (MAI), and hydrological perspective using groundwater drought index. For the analysis of SPI, rainfall data from 1981 to 2021 from Climate Hazards Group InfraRed Precipitation with Station data (CHIRPS) with the spatial resolution of 0.05° has been used. In addition, for the calculation of VCI, the NDVI band of Moderate Resolution Imaging Spectroradiometer (MODIS) (MOD13A2) datasets with 500 m pixel resolution from 2001 to 2021 have been used. The NDVI data of the previous 10 years is required to calculate VCI. Apart from that evapotranspiration (ET) band and potential evapotranspiration (PET) band of MODIS data (MOD16A2) of 2010–2011 and 2020–2021 with 500 m pixel resolution have been employed to calculate the moisture adequacy index (MAI). In addition, seasonal groundwater level data from 2000 to 2011 and 2010 to 2021 has been applied to calculate the groundwater drought index (GWDI) for the years 2010 and 2020, respectively. The groundwater level data has been collected from the CGWB, Kolkata, Government of India.

5.3.2 Methodology

5.3.2.1 Standardized Precipitation Index (SPI)

SPI represents the actual rainfall concerning the standardized departure of rainfall probability function (GoI., 2016). The SPI is an efficient and widely used technique to monitor drought conditions in terms of meteorological indices. For the computation of SPI, a minimum of 30 historical years of rainfall data is required. The negative value of SPI indicates dry conditions, whereas the positive value of the SPI denotes wet conditions. In this study, SPI has been calculated using Eq. 5.1:

$$\mathrm{SPI} = \frac{P_{ijk} - \overline{P}_{ij}}{\sigma_{ij}} \tag{5.1}$$

where P_{ijk} is the precipitation of individual year, $\overline{P}_{ij}$ is the mean precipitation of the historical past, and σ_{ij} is the standard deviation.

5.3.2.2 Moisture Adequacy Index (MAI)

MAI is used to determine the availability of moisture in the soil, which indicates the drought condition in terms of crop situation (GoI., 2016). The AET and PET have been implemented to compute the MAI for drought monitoring using Eq. 5.2. The higher value of MAI represents the less drought condition and the lower value of MAI denotes severe drought condition.

$$\mathrm{MAI} = \frac{\mathrm{AET}}{\mathrm{PET}} \times 100 \tag{5.2}$$

where AET is the actual evapotranspiration value and PET is the potential evapotranspiration value.

5.3.2.3 Vegetation Condition Index (VCI)

VCI is used to determine the drought condition in terms of vegetation health conditions. In this study, VCI has been computed based on the previous three-year NDVI value following Eq. 5.3 (GoI., 2016). The higher value of VCI indicates the good health of the vegetation cover, while the low value of the VCI reveals the poor condition of the vegetation health. Therefore, a low value of VCI is associated with drought conditions and vice versa.

$$\mathrm{VCI} = \frac{NDVI_i - NDVI_{\min}}{NDVI_{\max} - NDVI_{\min}} \tag{5.3}$$

where $NDVI_i$ denotes the NDVI value of the individual year, $NDVI_{\min}$ indicates the minimum NDVI value of the previous three years, and $NDVI_{\max}$ represents the maximum NDVI value of the previous three years.

5.3.2.4 Groundwater Drought Index (GWDI)

GWDI is a significant index to measure the drought condition in terms of the hydrological drought index. It is crucial for the assessment of groundwater availability for drinking and irrigation purposes. In the computation of GWDI, a minimum groundwater data of 10 observation years is required (GoI., 2016). The lower value of GWDI is associated with the deficit condition of groundwater and the higher value denotes the surplus condition of groundwater. The GWDI has been implemented to trace the hydrological drought condition in this study area using Eq. 5.4:

$$\mathrm{GWDI} = \frac{MGWD_j - GWD_{ij}}{GWD_{iMax}} \tag{5.4}$$

where $MGWD_j$ is the mean depth of the groundwater table below the surface (m), GWD_{ij} is the depth of the groundwater table in *i*th month and *j*th year (m), and GWD_{iMax} is the maximum depth to groundwater table in ith month in available dataset for *n* number of years.

5.3.3 Pattern of Drought and Related Processes

The anticipated climate change would likely worsen climate extremes like drought by further altering global patterns of precipitation and evaporation (Karl et al., 1997). Therefore, water security in terms of both quantity and quality is a major problem on a global scale. Droughts occur all around the world, although certain regions are more vulnerable than others (Smakhtin & Schipper, 2008). In the Indian context, the severity of drought varies across the agro-climatic regions. Few areas in the Ganga delta and its surrounding regions, especially the Damodar Fan Delta, are identified as critical and semi-critical regions in terms of the availability of groundwater resources. The recurring drought in this region has intensified the crisis of water resources. Thus, analysing the spatial distribution of the frequency, intensity, and duration of the DFD is necessary for overcoming the challenge of water crisis through drought monitoring and water resource management. The drought phenomena in this region have been characterized through the meteorological indicator, vegetation condition indicator, soil moisture indicator, and hydrological indicator using SPI, VCI, MAI, and GWDI, respectively, for the year 2010–2011 and 2020–2021, as well as for the pre-monsoon, monsoon, post-monsoon *rabi*, and post-monsoon *kharif* seasons. Besides, the spatial pattern of each index across the study region has been outlined meticulously.

The spatial distribution of SPI for 2010 and its seasonality are revealed in Fig. 5.5a–d. The SPI value is categorized into eight classes such as (1) extremely dry (<−2), (2) severely dry (−1.99 to −1.5), (3) moderately dry (−1.49 to −1), (4) mildly dry (−0.99 to 0), (5) mildly wet (0 to 0.99), (6) moderately wet (1–1.49), (7) severely wet (1.5–1.99), and (8) extremely wet (>2) condition. Regarding pre-monsoon season, the SPI reveals the normal condition with its value ranging from −0.09 to 0.05 (Fig. 5.5a). In this season, the majority of the study area (3062.80 km^2) falls under mildly dry condition. Besides, 145.11 km^2 has been characterized by mildly wet conditions, which is located eastern part of the study area. In the monsoon season, the SPI value differs from −0.24 to −0.45 which falls under the mildly dry condition throughout the study area (Fig. 5.5b). In the case of post-monsoon rabi season, the SPI value comprises between −0.14 and 0.02 which denotes the mildly dry to the mildly wet condition of drought. In this season, 3134.41 km^2 area is covered under the mildly dry condition and 73.5 km^2 area falls under the mildly wet condition (8.5c). In addition, the SPI value of the post-monsoon *kharif* season ranges from −0.27 to −0.11 indicating the mildly dry condition of the study area (8.5d). The spatial distribution of SPI of all seasons of 2010–2011 depicts that the mildly dry condition predominantly covers the upper part of the DFD, while the

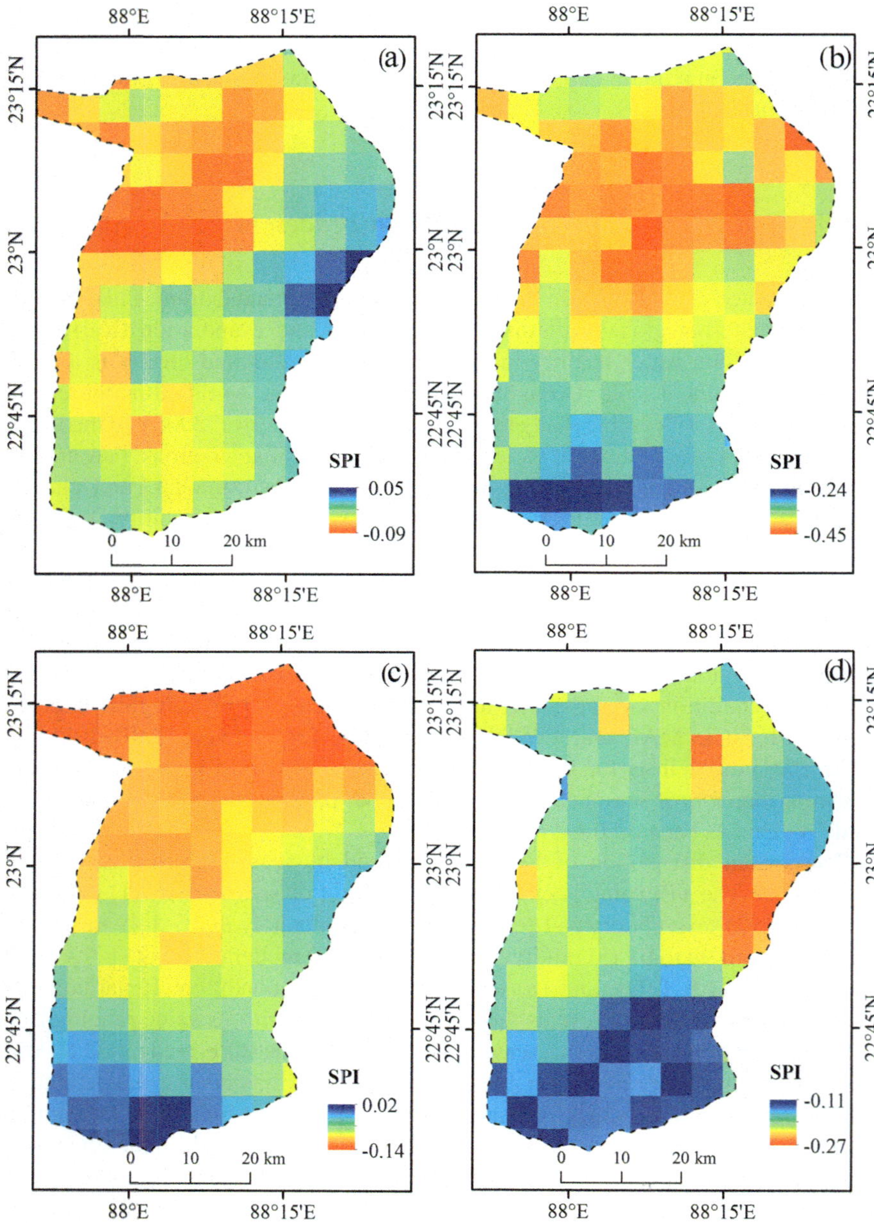

Fig. 5.5 Spatial distribution of SPI of 2010: (**a**) Pre-monsoon, (**b**) monsoon, (**c**) post-monsoon *rabi*, and (**d**) post-monsoon *kharif* seasons

lower part of the study area comprises relatively wet conditions (mildly wet) compared to the upper part.

In the year 2020–2021, the SPI value illustrated that the pre-monsoon period is characterized by mild wetness (0.48–0.73) (8.6a). The monsoon season is mildly dry to mildly wet condition (−0.15 to 0.02). In this season, a mildly wet class has been found in the southwestern part of the study area (Fig. 5.6b). Besides, the post-monsoon *rabi* season comprises the SPI value −0.2 to −1.2 which falls under the mildly dry condition (Fig. 5.6c). Similarly, the post-monsoon *kharif* season is considered a mildly dry condition in terms of meteorological drought comprising the SPI value −0.45 to −0.33 (Fig. 5.6d). Another noteworthy finding is that the upper part of the DFD consistently exhibits a dry region with negative SPI in comparison to the lower part of the basin except during the post-monsoon *kharif* season.

Additionally, the MAI has been employed for identifying the spatio-temporality of water-stressed conditions in the study region. In this study, MAI value has been categorized into four classes, including no drought (75–100), mild drought (50–75), moderate drought (25–50), and severe drought (0–25). The seasonality of MAI for 2010–2011 is portrayed that the pre-monsoon and post-monsoon *kharif* seasons contributed lion's share of severe drought conditions in the study region. Besides, the mild drought region is dominated in the monsoon season of this year. Besides, moderate drought is found to dominate in the post-monsoon *rabi* region (Fig. 5.7a–d; Table 5.1).

Moreover, regarding the MAI of 2020–2021, the pre-monsoon exhibits that the upper part of the DFD comes under a severe drought-prone area (1638.93 km^2), whereas the lower part (1533 km^2) displays a moderate drought region (Fig. 5.8a). While shifting to the monsoon season, the absence of severe and moderate drought is observed. Additionally, 887.81 km^2 of the region is found as mild drought, and 1997.25 km^2 has no drought condition in the monsoon season. Moreover, the post-monsoon *rabi* represents the dominant mild drought (2207.46 km^2) followed by moderate drought (961.4 km^2) (Fig. 5.8c; Table 5.1). Among all the seasons, the severity of the drought has been observed for the post-monsoon *kharif* season where 3039.69 km^2 of the study region comes under the severe drought-prone region. Therefore, the study area experiences drought conditions of different magnitudes in four seasons. It is also found that severe drought covers the majority of the study area. In terms of moderate to no drought conditions, these were mostly found in the lower part of the study area.

With a similar time and seasonal frame, the VCI has been measured to signal the effect of meteorological conditions directly on the health of vegetation. In terms of drought conditions, the VCI value has been grouped into three classes, such as good (0.6–1), fair (0.4–0.6), and poor (0–0.4) condition of vegetation health. The poor condition of vegetation health is associated with drought events and vice versa. The year 2010–2011 exhibits the changes in VCI in different seasons.

For example, the pre-monsoon season revealed that vegetation covers with fair (374.22 km^2) and poor (74.18 km^2) conditions in a few areas, especially located in the upper part of the study area (Fig. 5.9a). Besides, the 2759.51 km^2 area falls under the good condition of VCI. The monsoon season portrays the normal

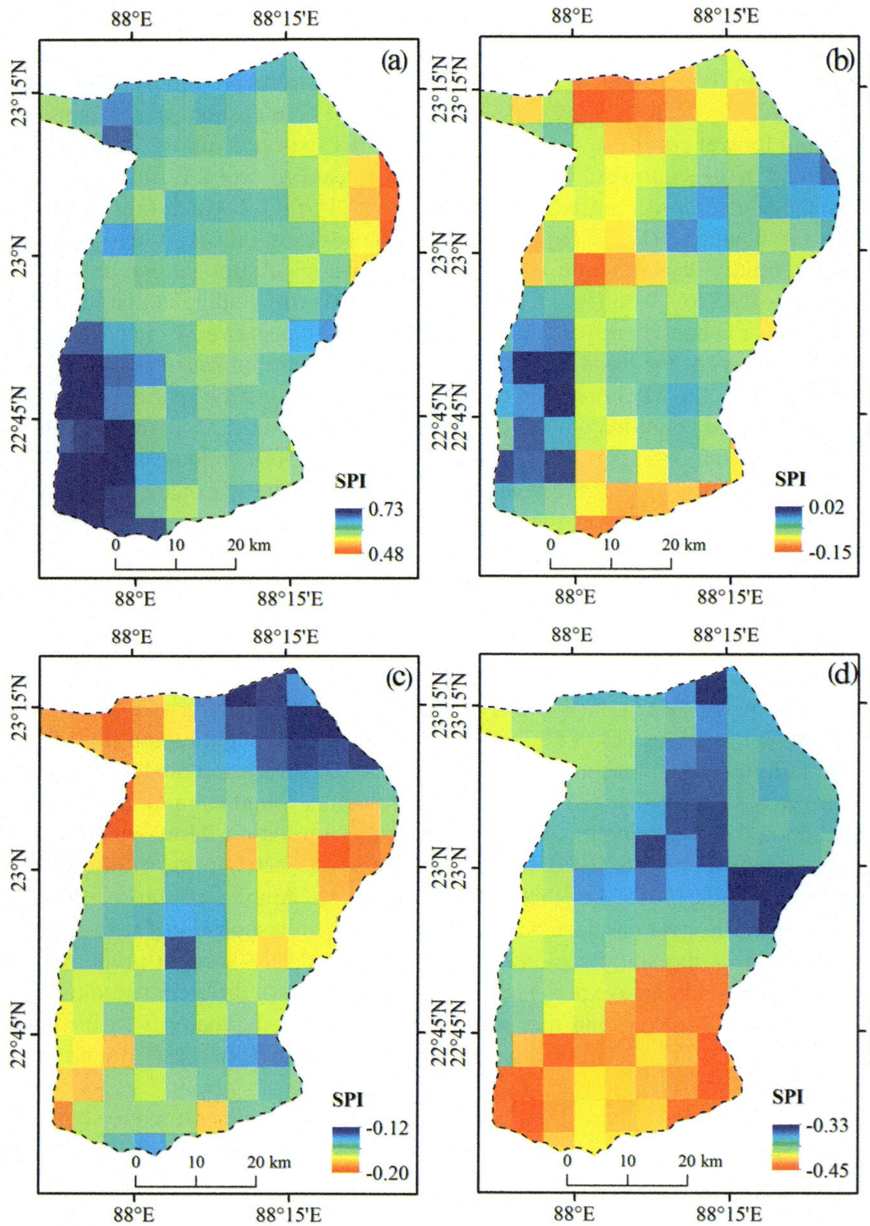

Fig. 5.6 Spatial distribution of SPI of 2020: (**a**) Pre-monsoon, (**b**) monsoon, (**c**) post-monsoon *rabi,* and (**d**) post-monsoon *kharif* seasons

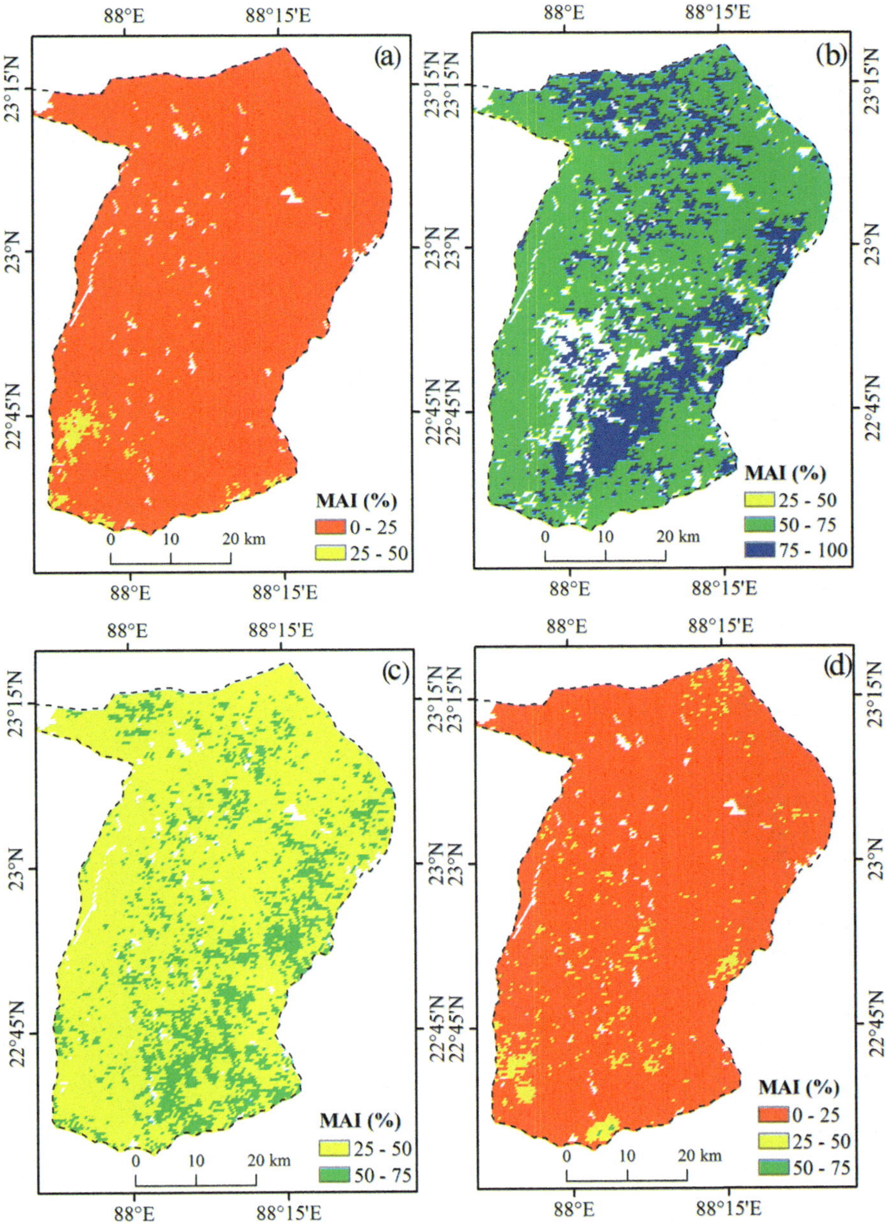

Fig. 5.7 Spatial distribution of MAI of 2010: (**a**) Pre-monsoon, (**b**) monsoon, (**c**) post-monsoon *rabi*, and (**d**) post-monsoon *kharif* seasons

Table 5.1 Distribution of drought-affected area

Drought indicators	Classes	Ranges	Area (km^2) in 2010–2011				Area (km^2) in 2020–2021			
			Pre-monsoon	Monsoon	Post-monsoon rabi	Post-monsoon kharif	Pre-monsoon	Monsoon	Post-monsoon rabi	Post-monsoon kharif
SPI	Extremely dry	<−2	0.00	0.00	0.00	0.00	0.00	0.00	0.00	0.00
	Severely dry	−1.99 to −1.5	0.00	0.00	0.00	0.00	0.00	0.00	0.00	0.00
	Moderately dry	−1.49 to −1	0.00	0.00	0.00	0.00	0.00	0.00	0.00	0.00
	Mildly dry	−0.99 to 0	3062.80	3207.91	3134.41	3207.91	0.00	3122.95	3207.91	3207.91
	Mildly wet	0–0.99	145.11	0.00	73.50	0.00	3207.91	84.96	0.00	0.00
	Moderately wet	1–1.49	0.00	0.00	0.00	0.00	0.00	0.00	0.00	0.00
	Severely wet	1.5–2	0.00	0.00	0.00	0.00	0.00	0.00	0.00	0.00
	Extremely wet	>2	0.00	0.00	0.00	0.00	0.00	0.00	0.00	0.00
MAI	No drought	76–100	0.00	742.79	0.00	0.00	0.00	1997.25	3.39	0.00
	Mild drought	51–75	0.00	2152.49	652.49	1.09	0.55	887.81	2207.46	1.07
	Moderately drought	26–50	62.33	6.96	2519.77	147.71	1533.00	0.00	961.40	139.93
	Severe drought	0–25	3110.28	0.00	0.00	3023.88	1638.93	0.00	0.00	3031.69
VCI	Good	0.6–1	2759.51	3201.05	223.02	1351.25	2184.01	3119.78	133.57	714.60
	Fair	0.4–0.6	374.22	6.86	445.79	682.15	576.99	59.82	342.77	509.88
	Poor	0–0.4	74.18	0.00	2539.11	1174.51	446.91	28.31	2731.57	1983.44

GWDI	Normal	> − 0.15	2635.52	476.62	607.04	2197.09	3134.17	2390.59	2139.43	3207.91
	Mild	−0.16 to −0.3	482.70	687.56	1852.71	848.09	53.92	443.60	355.77	0.00
	Moderate	−0.31 to −0.45	61.05	1087.68	652.93	106.93	12.85	222.35	217.31	0.00
	Severe	−0.46 to −0.6	19.71	707.66	87.10	39.09	6.89	84.15	174.89	0.00
	Extreme	<−0.6	8.93	248.39	8.13	16.72	0.08	67.22	320.51	0.00

Sources: Computed by the authors, 2022

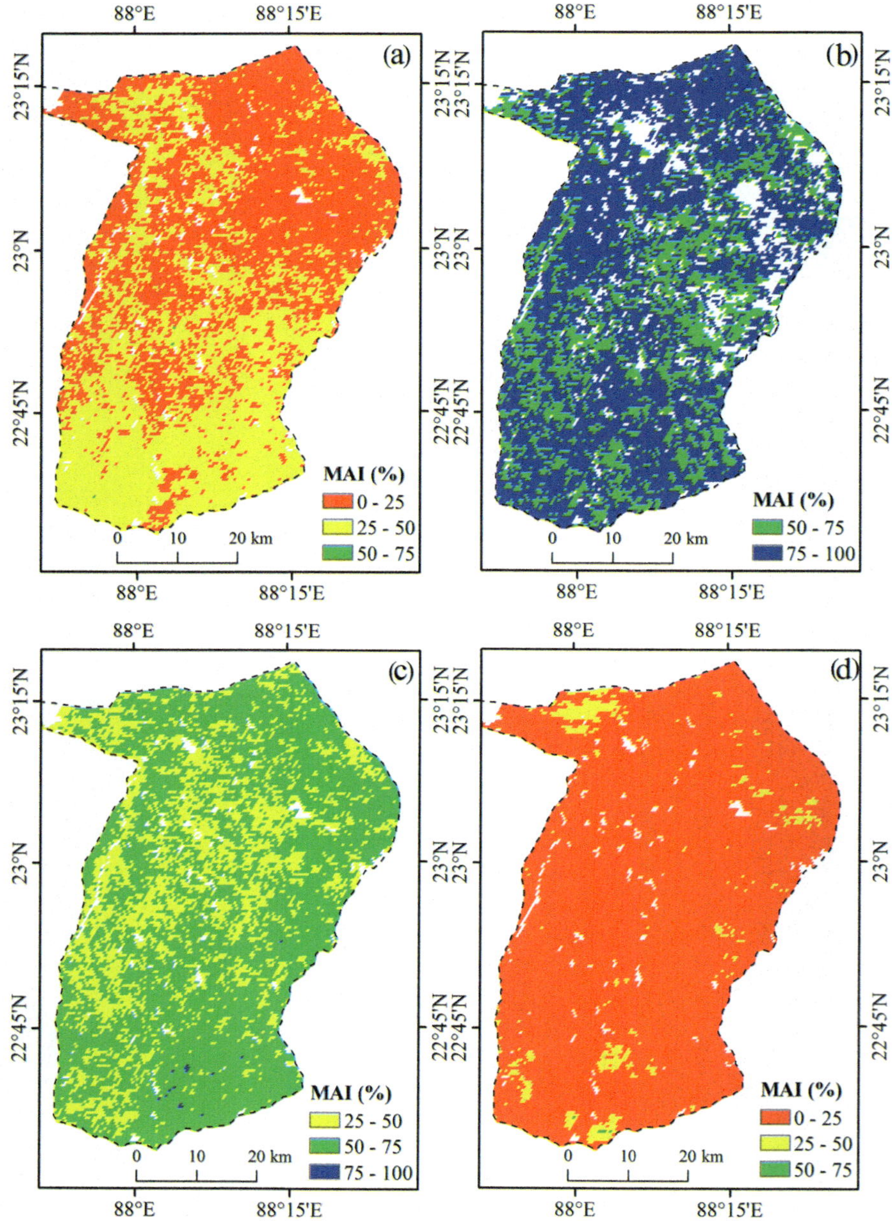

Fig. 5.8 Spatial distribution of MAI of 2020: (**a**) Pre-monsoon, (**b**) monsoon, (**c**) post-monsoon *rabi*, and (**d**) post-monsoon *kharif* seasons

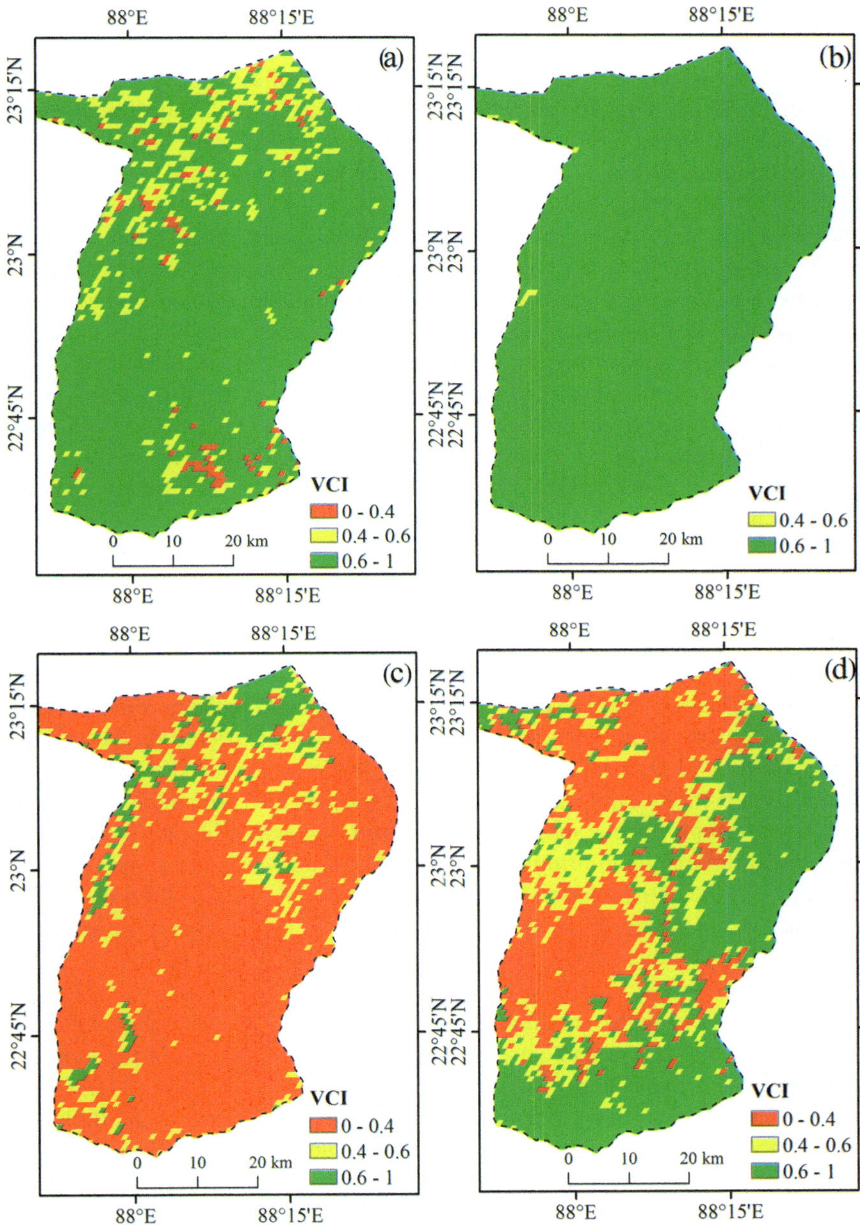

Fig. 5.9 Spatial distribution of VCI of 2010: (**a**) Pre-monsoon, (**b**) monsoon, (**c**) post-monsoon *rabi,* and (**d**) post-monsoon *kharif* seasons

condition for the entire region except for few locations with fair vegetation health conditions (Fig. 5.9b). However, it is observed that with the beginning of post-monsoon *rabi* season, the VCI has started decreasing for the majority of the region (Fig. 5.9c), indicating that the area falls under poor vegetation health conditions. It also depicts that in the post-monsoon *kha*rif season, 1174.51 km^2 area falls under poor vegetation health conditions and 682.15 km^2 area is comprised by the fair condition of VCI (Fig. 5.9d; Table 5.1).

Moreover, in this season, the good VCI class has been found in the south and eastern part of the study area comprising the areal extension of 1351.25 km^2. Therefore, a general trend has been observed here that the pre-monsoon and monsoon seasons portray the normal condition, while the transformation from normal to severe conditions is encountered for the rest of the seasons. Besides, the VCI of 2020–2021 follows the same pattern, but a completely antagonistic scenario has been observed. In the pre-monsoon season, the good vegetation condition comprises 2184.04 km^2, which is found in the southern and northern part of the study area. In this season, fair and poor class of VCI have been observed in the northern and middle part of the study area comprising 576.99 and 446.91 km^2, respectively (8.10a). The vegetation health condition in the monsoon season is predominantly comprised of the good class of VCI (8.10b). Besides, in the post-monsoon *rabi* season, the majority of the study area is covered by poor VCI classes except for a few locations in the western part (Fig. 5.10c). Moreover, in the post-monsoon *kharif* season, the good and fair class of VCI covers the spatial extension of 714.60 and 509.88 km^2, respectively. The good and fair class of VCI has been found in the northern, eastern, and southern parts of the study area (Fig. 5.10d; Table 5.1). Therefore, the seasonal analysis of the VCI signifies that the study area is a water-stressed region, particularly in the post-monsoon season. The crisis of water resources is therefore an indication that this region needs special attention from the perspective of water resource management through monitoring the surface and groundwater resources.

Apart from the above-motioned three indices, GWDI has also been considered for characterizing the spatial and seasonal distribution of groundwater-stressed conditions. The GWDI has been grouped into five classes such as normal (>−0.15), mild (−0.15 to −0.3), moderate (−0.3 to −0.45), severe (−0.45 to −0.6), and extreme (<−0.6) drought conditions. In 2010–2011, the spatial distribution of GWDI reveals that severe and extreme drought conditions appear in the monsoon season. It is also found that the severe (707.66 km^2) and extreme (248.39 km^2) drought conditions are found in the eastern and northeastern parts of the study area (Fig. 5.11b). In the case of pre-monsoon season, the dominant part (2635.52 km^2) of the study area is comprised of no drought conditions (Fig. 5.11a). Similarly, the post-monsoon *kharif* season comprises no drought conditions in the major part (2197.09 km^2) of the study area (Fig. 5.11d). Additionally, in the post-monsoon *rabi* season, mild drought is observed predominantly (1852.71 km^2) in the study area (Fig. 5.11c).

Similarly, in 2020–2021, severe and extreme drought are found in the monsoon and post-monsoon rabi season, and pre-monsoon and post-monsoon *kharif* season are characterized by mild to no drought conditions (Fig. 5.12a–d).

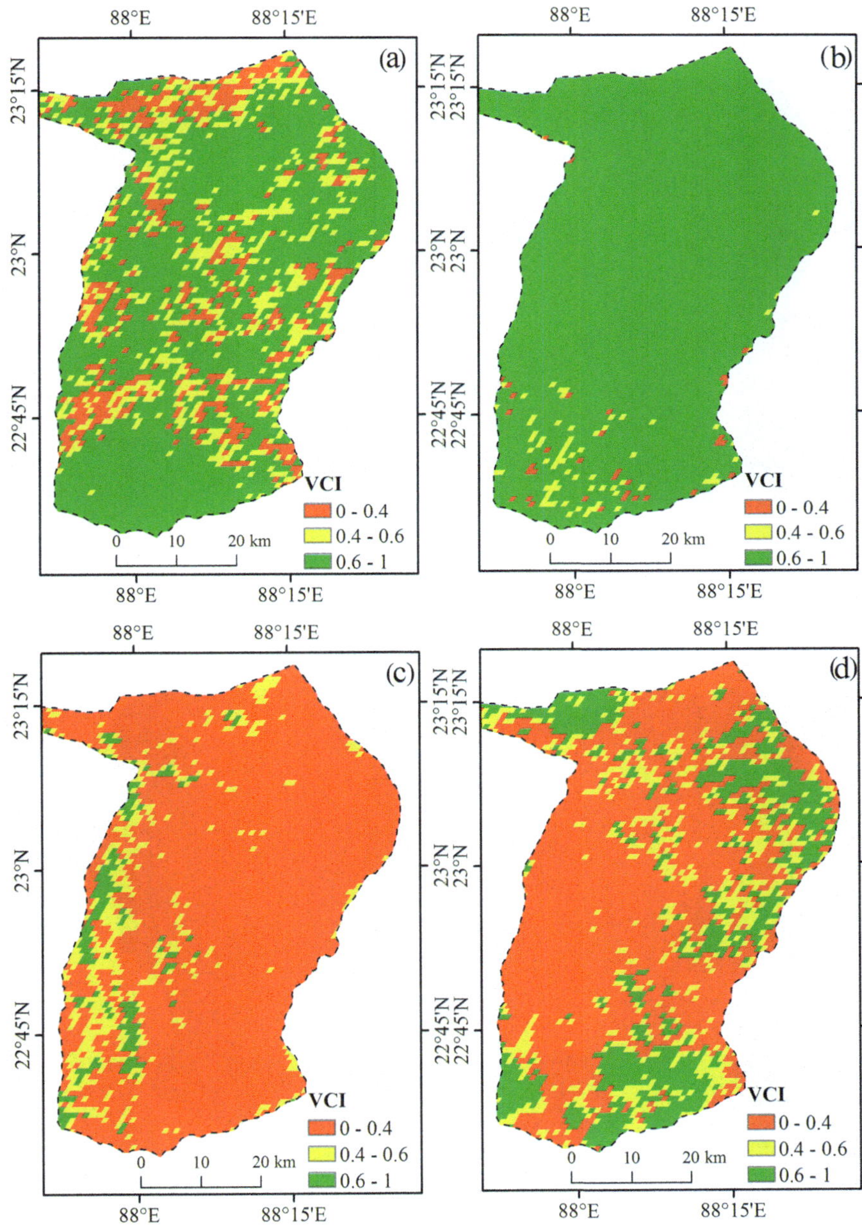

Fig. 5.10 Spatial distribution of VCI of 2020: (**a**) Pre-monsoon, (**b**) monsoon, (**c**) post-monsoon *rabi*, and (**d**) post-monsoon *kharif* seasons

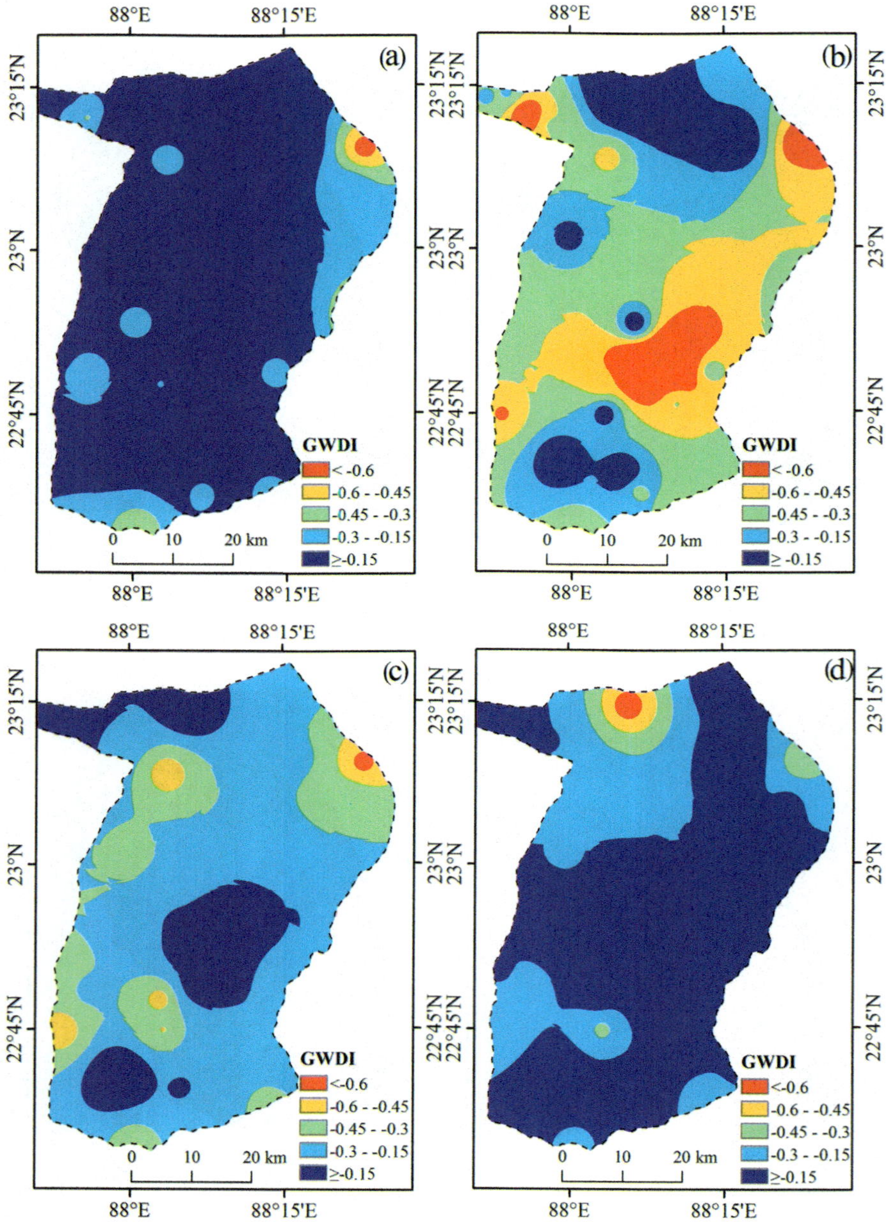

Fig. 5.11 Spatial distribution of GWDI of 2010: (**a**) Pre-monsoon, (**b**) monsoon, (**c**) post-monsoon *rabi*, and (**d**) post-monsoon *kharif* seasons

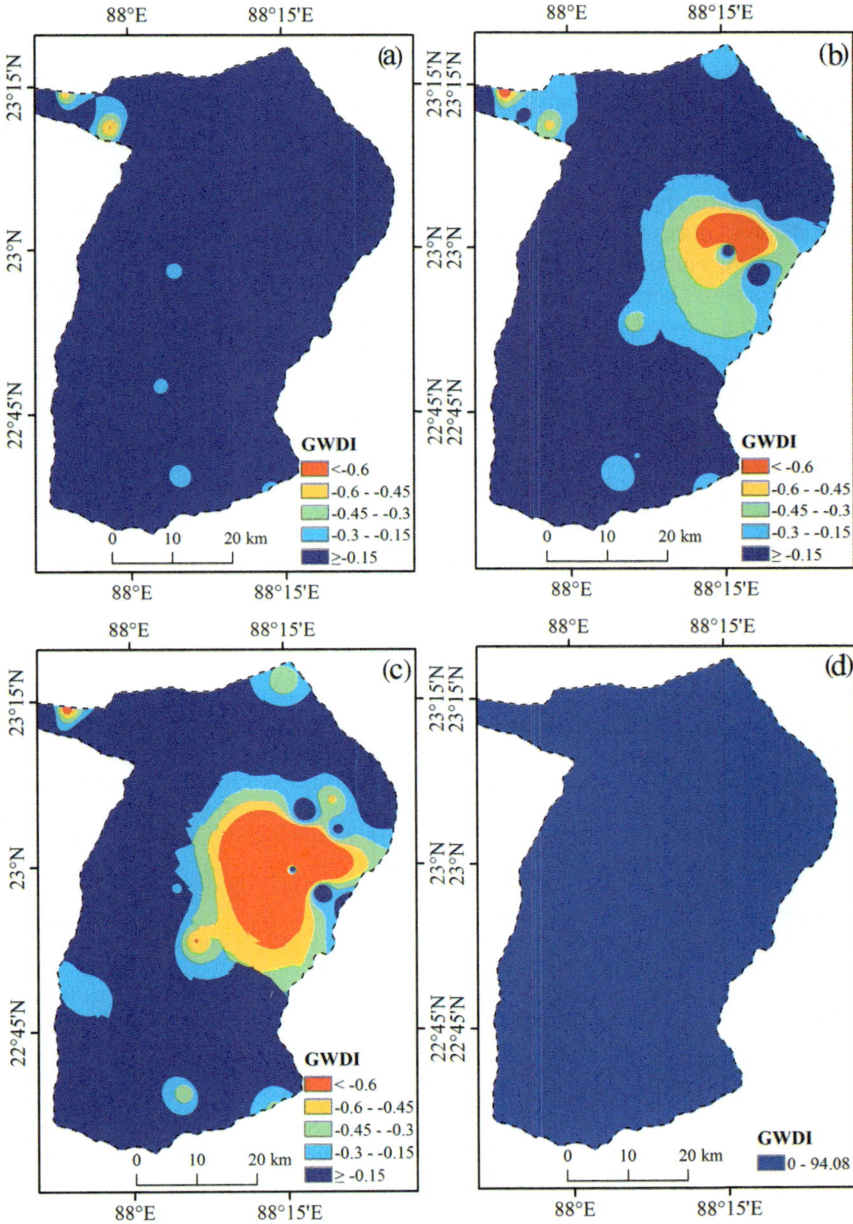

Fig. 5.12 Spatial distribution of GWDI of 2020: (**a**) Pre-monsoon, (**b**) monsoon, (**c**) post-monsoon *rabi,* and (**d**) post-monsoon *kharif* seasons

In the monsoon season, the severe and extreme drought comprise the areal extension of 84.15 and 67.22 km^2, which are found in the eastern part of the study area (Fig. 5.12b). Similarly, in the post-monsoon rabi season, the severe and extreme drought accounts for 174.89 and 320.51 km^2, respectively (Fig. 5.12c; Table 5.1). Besides, according to GWDI, extreme and severe droughts are prominently found during monsoon season, while they are less representative of the other season. This particular pattern appears due to the reason that during the monsoon season, the region receives maximum rainfall which has a substantial effect on augmenting the groundwater resources of the following seasons, most particularly in the post-monsoon rabi season. Therefore, the acute deficit in groundwater resources has not been observed in post-monsoon season, although the post-monsoon season receives the less amount of rainfall. Additionally, it should be mentioned that the anthropogenic pressure on groundwater coupled with less rainfall in the post-monsoon and pre-monsoon seasons again leads to the deficit in groundwater resources in the monsoon season.

Therefore, an attempt has been made in Chap. 6 to showcase the role of the palaeochannels in the water resource management using the flood surplus and flood deficit condition of the DFD.

References

Agnihotri, A. K., Ohri, A., Gaur, S., Shivam, Das, N., & Mishra, S. (2019). Flood inundation mapping and monitoring using SAR data and its impact on Ramganga River in Ganga basin. *Environmental Monitoring and Assessment, 191*. https://doi.org/10.1007/s10661-019-7903-4

Allegretti, G., Montoya, M. A., & Talamini, E. (2022). Assessing sectoral water stress states from the demand-side perspective through water footprint dimensions decomposition. *Science of the Total Environment, 809*, 152216.

Bhattacharyya, K. (2011). *The lower Damodar River, India: Understanding the human role in changing fluvial environment*. Springer.

Boretti, A., & Rosa, L. (2019). Reassessing the projections of the world water development report. *npj Clean Water, 2*. https://doi.org/10.1038/s41545-019-0039-9

Das, B., Pal, S. C., & Malik, S. (2017). Assessment of flood hazard in a riverine tract between Damodar and Dwarkeswar River, Hugli District, West Bengal, India. *Spatial Information Research, 26*, 91–101. https://doi.org/10.1007/s41324-017-0157-8

Falkenmark, M. (1989). The massive water scarcity now threatening Africa: Why isn't it being addressed? *Ambio*, 112–118.

Ghosh, S., & Mistri, B. (2012). Hydrogeomorphic significance of sinuosity index in relation to river instability: A case study of Damodar River, West Bengal, India. *International Journal of Advances in Earth Sciences, 1*, 49–57.

Ghosh, S., Hoque, M. M., Islam, A., Barman, S. D., Mahammad, S., Rahman, A., & Maji, N. K. (2022). Characterizing floods and reviewing flood management strategies for better community resilience in a tropical river basin, India. Natural Hazards. https://doi.org/10.1007/s11069-022-05618-y.

GoI (2016). Manual for drought management. Department of Agriculture, Cooperation & Farmers Welfare Ministry of Agriculture & Farmers Welfare Government of India New Delhi, Retrieved January 25, 2022, from https://agricoop.nic.in/sites/default/files/Manual%20Drought%20 2016.pdf

Hertel, T., & Liu, J. (2019). Implications of water scarcity for economic growth. In *Economy-wide modeling of water at regional and global scales* (pp. 11–35). Springer Singapore. https://doi.org/10.1007/978-981-13-6101-2_2

Howe, C. W. (1976). The effects of water resource development on economic growth the conditions for success. *Natural Resources Journal, 16*, 939–955.

Irrigation and Waterways Directorate. (2015). *Annual flood report for the year 2015*. Advance Planning, Project Evaluation & Monitoring Cell.

Irrigation and Waterways Directorate. (2016). *Annual flood report 2016*. Advance Planning, Project Evaluation & Monitoring Cell.

Irrigation and Waterways Directorate. (2017). *Annual flood report for the year 2017*. Advance Planning, Project Evaluation & Monitoring Cell.

Irrigation and Waterways Directorate. (2019). *Annual flood report 2019*. Advance Planning, Project Evaluation & Monitoring Cell.

Karl, T. R., Nicholls, N., & Gregory, J. (1997). The coming climate. *Scientific American, 276*(5), 78–83.

Kumar, M. D., Shah, Z., Mukherjee, S., Mudgerikar, A., et al. (2008). *Water, human development and economic growth: Some international perspectives. IWMI-Tata water policy research program's seventh annual partners' meet*. ICRISAT.

Mahammad, S., & Islam, A. (2021). Evaluating the groundwater quality of Damodar Fan Delta (India) using fuzzy-AHP MCDM technique. *Applied Water Science, 11*. https://doi.org/10.1007/s13201-021-01408-2

Melkamu, T., Bagyaraj, M., Adimaw, M., Ngusie, A., & Karuppannan, S. (2022). Detecting and mapping flood inundation areas in Fogera-Dera Floodplain, Ethiopia during an extreme wet season using Sentinel-1 data. *Physics and Chemistry of the Earth, Parts A/B/C, 127*, 103189. https://doi.org/10.1016/j.pce.2022.103189

Mirdashtvan, M., Najafinejad, A., Malekian, A., & Sa'doddin, A. (2021). Sustainable water supply and demand management in semi-arid regions: Optimizing water resources allocation based on RCPs Scenarios. *Water Resour Manage, 35*, 5307–5324. https://doi.org/10.1007/s11269-021-03004-0

Moharrami, M., Javanbakht, M., & Attarchi, S. (2021). Automatic flood detection using sentinel-1 images on the google earth engine. *Environmental Monitoring and Assessment, 193*. https://doi.org/10.1007/s10661-021-09037-7

Rudra, K., Rudra, L., & Leuchner. (2018). *Rivers of the ganga-Brahmaputra-Meghna Delta*. Springer.

Smakhtin, V. U., & Schipper, E. L. (2008). Droughts: The impact of semantics and perceptions. *Water Policy, 10*, 131–143. https://doi.org/10.2166/wp.2008.036

Tiwari, V., Kumar, V., Matin, M. A., Thapa, A., Ellenburg, W. L., Gupta, N., & Thapa, S. (2020). Flood inundation mapping-Kerala 2018; harnessing the power of {SAR}, automatic threshold detection method and Google earth engine. (G. J.-P. Schumann, Ed.). *PLoS One, 15*, e0237324. https://doi.org/10.1371/journal.pone.0237324

Tripathi, G., Pandey, A. C., Parida, B. R., & Kumar, A. (2020). Flood inundation mapping and impact assessment using multi-temporal optical and SAR satellite data: A case study of 2017 flood in Darbhanga District, Bihar, India. *Water Resources Management, 34*, 1871–1892. https://doi.org/10.1007/s11269-020-02534-3

Usman, R., Mirzania, P., Alnaser, S. W., Hart, P., & Long, C. (2022). Systematic review of demand-side management strategies in power systems of developed and developing countries. *Energies, 15*(21), 7858.

Chapter 6
Water Resource Management Using Palaeochannels

6.1 Role of Palaeochannels in Groundwater and Surface Water Augmentation

Palaeochannels play an instrumental role in shaping the water resource management of the different areas of the world as revealed through several studies at the national and global scale. For example, Rathore et al. (2010) examined the drought-prone Rajasthan state for detecting the palaeochannels because these act as the potential aquifer in Mendha River catchment in Western India. Their study indicates that palaeochannel aquifer exhibits higher potential compared to the surrounding aquifers. Similarly, Kaur et al. (2024) identified palaeochannels in dry areas of Punjab for water security and attainment of the sustainable development of goals 6 and 15. This study found palaeochannel filled with loose, permeable sediments (e.g. sand and gravel) making them natural reservoirs capable of storing large volumes of groundwater. Thus, palaeochannels may be used for artificial groundwater recharge as demonstrated by Samadder et al. (2011) in the lower Ganga plains. Conversely, these palaeochannels if renovated can store excess floodwater.

This chapter deals with the potentiality of the palaeochannels in groundwater recharge, and the direction and dimensions of the recharge area. Finally, this chapter will concentrate on people's perceptions of the existing flood management strategies and future flood management with palaeochannels.

S. Mahammad, A. Islam, *Palaeochannels and Water Resource Management*,
https://doi.org/10.1007/978-3-031-92021-9_6

6.1.1 Volumetric Augmentation of Water Resources

Palaeochannels play a crucial role in the augmentation of the surface and groundwater both in terms of the volumetric (quantitative) increase and qualitative improvement. Regarding the volumetric increase in groundwater, it is found that the depth of the groundwater during the observation period 2013–2014 was 8.96 mbgl for the palaeochannel area, while it was 14.18 mbgl for the non-palaeochannel area (Table 6.1). This implies that groundwater is located much deeper in the non-palaeochannel area. This may be due to the higher rate of recharge of the groundwater through the palaeochannels. This trend of the pre-monsoon season is most notable because this is the lean season where the water level drops quickly. Apart from the average scenario of the water level, the variations are higher for the well located in the non-palaeochannel area compared to the palaeochannel area implying the relative consistency of the palaeochannel in the groundwater recharge. The F-test also shows that there is a significant variation in the water level between the palaeochannel and non-palaeochannel area (Table 6.1). Similar to the pre-monsoon seasons, the other seasons also portray a similar picture with the varying magnitude of the water levels. The difference between water level in the palaeochannel and non-palaeochannel area is also greater for the post-monsoon *kharif* season followed by the post-monsoon *rabi* and monsoon seasons. The water level difference between the palaeochannel wells and non-palaeochannel wells is comparatively lower because of the monsoon spell and its consequences. However, in general, the water level is always lower for the palaeochannel wells than the non-palaeochannel wells implying the efficiency of the palaeochannel aquifer to pass the surface water through layers of the palaeochannel wells. It is also observed that with the passage of time, the water depth has fallen in both the palaeochannel and non-palaeochannel areas. Thus, it is observed that there is a higher rate increase in the water depth for the non-palaeochannel wells compared to the palaeochannel wells (Table 6.1).

Regarding the surface water dynamics, a buffer of 1 km from the centerline of the palaeochannel was drawn. The inner buffer (0–1 km) and outer buffer (1–2 km) portray the areal dynamics of the surface water bodies in the pre-monsoon and post-monsoon seasons. It is found that there is a higher concentration of water bodies in the inner buffer (0–1 km) compared to the outer buffer (1–2 km). It is observed that pre-monsoon season water bodies are less compared to the post-monsoon area due to the effects of the monsoonal rainfall in the monsoon of July–September. The F-statistic also portrays that there is a significant difference between the palaeochannel and non-palaeochannel areas regarding the presence of surface water bodies because it is found that the palaeochannel areas are more common in the inner circle of the buffer and decline outward. Thus, it proved that the expansion of the surface waterbody is higher due to the presence of the palaeochannels (Table 6.2).

Table 6.1 Groundwater dynamics

2013–2014								
Season	**Pre-monsoon**		**Monsoon**		**Post-monsoon *rabi***		**Post-monsoon *kharif***	
Types	**Palaeochannel**	**Non-palaeochannel**	**Palaeochannel**	**Non-palaeochannel**	**Palaeochannel**	**Non-palaeochannel**	**Palaeochannel**	**Non-palaeochannel**
Average	8.96	14.18	7.07	10.51	5.35	8.31	6.49	10.98
SD	2.33	3.34	2.85	3.33	1.51	3.07	2.26	3.42
F	13.92		5.15		5.63		9.96	
P-value	0.001		0.032		0.025		0.004	
F crit*	4.20		4.23		4.20		4.20	
2020–2021								
Seasons	**Pre-monsoon**		**Monsoon**		**Post-monsoon *rabi***		**Post-monsoon *kharif***	
Types	**Palaeochannel**	**Non-palaeochannel**	**Palaeochannel**	**Non-palaeochannel**	**Palaeochannel**	**Non-palaeochannel**	**Palaeochannel**	**Non-palaeochannel**
Average	9.65	15.88	8.10	12.30	7.75	11.69	9.72	14.21
SD	2.50	3.17	2.08	3.10	2.42	3.48	2.91	3.00
F	21.17		10.54		7.33		11.37	
P-value	0.000		0.003		0.011		0.002	
F crit*	4.20		4.20		4.20		4.20	

Sources: Computed by the researcher, 2022 (Note: * significance level 0.05, degree of freedom (df) = 1)

Table 6.2 Surface water dynamics

Season	Pre-monsoon (waterbody in km^2)		Post-monsoon (waterbody in km^2)	
Buffer from palaeochannel	Buffer 1 (1 km)	Buffer 2 (2 km)	Buffer 1 (1 km)	Buffer 2 (2 km)
Average	48.35	15.29	53.22	17.41
SD	2.80	0.71	3.25	0.86
F	394.06		340.47	
P-value	0.000001		0.000002	
F crit*	5.99		5.99	

Sources: Computed by the researcher, 2022 (Note: * significance level 0.05, degree of freedom (df) = 1)

6.1.2 Augmentation of Water Quality

Water quality improvement in the palaeochannel is also noted in the previous literature. In this study, water quality for drinking purposes (WQI) and water quality for irrigation purposes (EIWQI) were tested separately for both the groundwater and surface water. In this chapter, an attempt has been made to link water quality with palaeochannel. It is found that irrigation and drinking water quality is better for the groundwater wells located in the palaeochannel compared to the non-palaeochannel wells. For example, WQI for palaeochannel wells is 45.60 (groundwater) and 68.12 (surface water), while it is 72.80 (groundwater) and 75.66 (surface water) for non-palaeochannel wells. Similarly, EIWQI for palaeochannel wells is 42.77 (groundwater) and 71.10 (surface water), whereas it comprises 58.23 (groundwater) and 74.46 (surface water) for non-palaeochannel wells. This indicates that this difference in water quality is due to the presence of the sandy layers in the palaeochannel that reduces the pollution concentration. Moreover, it is also observed that palaeochannel is not always considered a water quality improvement tool. There are other dimensions of water quality including organic pollution and heavy metal contamination that may slightly increase in the palaeochannel areas. There are some studies (Sarkar & Islam, 2019; Kakade et al., 2021) which indicate that palaeochannel promotes the annual growth of plants at an accelerated rate that is supposed to induce organic pollution. It is the situation created due to the stagnation of water within the palaeochannel in most cases just like the condition in a eutrophied lake. However, the OIP may not be excessively high (1.62) for the palaeochannel than non-palaeochannel area (1.56). The F-test also shows that this difference in the statistical sense is not significant ($p = 0.51$) (Table 6.3). Moreover, heavy metal pollution also becomes higher due to the elevated concentration of iron and arsenic in these areas. It is found that palaeochannel areas are associated with arsenic contamination due to the oxidized environment created by the acclerated withdrawal of groundwater for *boro* cultivation (Pal & Mukherjee, 2008, 2010; Mahammad & Islam, 2021). Thus, water quality improvement is not linear like the volumetric increase in water.

Table 6.3 Groundwater and surface water quality

Parameters	Groundwater				Surface water			
	WQI		EIWQI		WQI		EIWQI	
Types	Palaeochannel	Non-palaeochannel	Palaeochannel	Non-palaeochannel	Palaeochannel	Non-palaeochannel	Palaeochannel	Non-palaeochannel
Mean	45.60	72.80	42.77	58.23	68.12	75.66	71.10	74.46
SD	15.87	26.36	13.50	33.36	11.08	9.31	25.51	17.30
F	10.64		2.50		3.18		0.13	
P-value	0.003		0.124		0.083		0.723	
F crit*	4.15		4.15		4.13		4.13	

Sources: Computed by the authors, 2022 (Note: *significance level 0.05, degree of freedom (df) = 1)

Although irrigation and drinking water quality may improve, organic pollution and heavy metal pollution may increase in some cases.

6.2 Integrated Water Resource Management Using Palaeochannel

Integrated water resource management entails the management of surface water and groundwater in such a way that reduces the vulnerability of floods and droughts. In this study, an attempt has been made to develop a framework for IWRM considering the geophysical conditions of the palaeochannel and socio-economic needs and demands as discussed in the following sections.

6.2.1 Geo-Physical Consideration of Water Resource Management through Palaeochannels

Palaeochannels are useful tools for water resource management, which has been discussed in Chap. 1 for their groundwater recharge potential on the one hand and diversion medium to excess floodwater on the other. For this reason, few researchers (Sinha et al., 2013; Samadder et al., 2011; Aslam et al., 2003; Chaudhary & Aggarwal, 2009; Rathore et al., 2010) have exhibited water resource management through palaeochannels. The DFD has immense potential for palaeochannel to water resource management. The locational pattern and direction of water recharge are also shown in Fig. 6.1. The potentiality for groundwater recharge has been analysed based on the recharge rate of the subsurface materials. For this purpose, the water table fluctuation (WTF) method proposed by the Groundwater Estimation Committee (GEC) in 1997 (Ministry of Water Resources Government of India, 2009) has been implemented. The WTF method is easy to apply and suitable for the unconfined aquifer system, especially in the region where monsoon rainfall is the main source of recharge water (CGWB, 2014). In this method, the recharge rate of groundwater is calculated based on the multiplication of specific yield (Sy) value with the seasonal fluctuation of groundwater level (dh/dt). This study area is located in the unconfined aquifer system and rainfall is the main source of recharge. Therefore, the potential of the groundwater recharge site has been delineated with the help of the recharge rate according to the WTF method. For this purpose, the Sy value is computed after Johnson (1967). The block-wise groundwater level for pre-monsoon and post-monsoon seasons has been computed from the State Water Investigation Directorate (SWID) Kolkata, Government of West Bengal. The result depicts that the gross recharge rate ranges from 0.69 to 217.22 with an average of 45.14 ± 23.77. The distribution of gross recharge rate has been classified into five

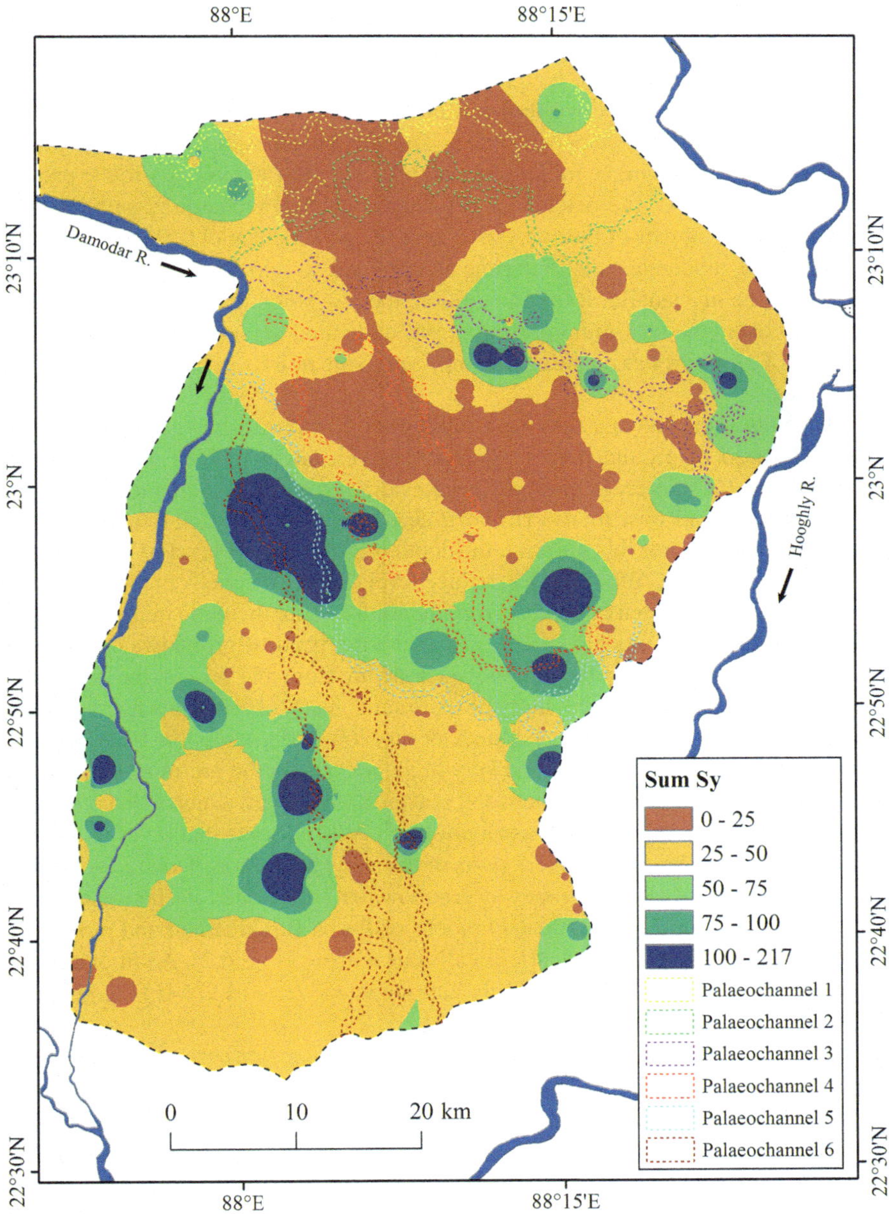

Fig. 6.1 Spatial distribution of the sum of the specific yield of the lithological logs

classes: very low (0–25), low (25–50), medium (50–75), high (75–100), and very high (100–217.22) for groundwater recharge potential (Fig. 6.1).

The spatial distribution of the gross recharge rate portrays that a very high class is comprised in the middle western and middle eastern part of the study area near the palaeochannel 4 and palaeochannel 5. The small patches of the very high potential zone are confined in the middle part of the study area and vicinity to the palaeochannel 2. Moreover, the patches of very high potential recharge zone have been found in the lower part of the study area near the palaeochannel 6. Apart from that, the high and moderate potential zone is confined to surrounding the very high potential zone in a smaller form. In addition, the low potential zone of groundwater recharge covers the maximum portion of the study area. Besides, a very low potential zone of groundwater recharge is comprised in the middle to the upper part of the study area.

The study also aims to assess the groundwater flow direction with the help of the hydraulic head of groundwater or groundwater elevation (GWE) with respect to mean sea level. The average groundwater elevation of pre-monsoon, monsoon, post-monsoon *rabi*, and post-monsoon *kharif* seasons from 2010–2011 to 2020–2021 has been analysed. For this purpose, the depth of groundwater level of 51 wells was collected from the CGWB. Besides, reduce level (RL) of the wells in terms of mean sea level (MSL) has been computed from the shuttle radar thematic mapper (SRTM) and digital elevation model (DEM). The hydraulic head of groundwater has been calculated by subtracting the groundwater depth from the RL of the respective wells. The results illustrated that the hydraulic head of groundwater has been grouped into five elevation classes, such as very low (<0 m), low (0–10 m), moderate (10–20 m), high (20–30 m), and very high (>30 m). In the pre-monsoon season, it is observed that GWE in pre-monsoon season is −10.39 m, which is the minimum among the seasons. The spatial distribution of GWE portrays that a very low class comprises the eastern and southern parts of the study area. Moreover, the very high, high, and moderate elevated zones of groundwater are situated in the western part of the study area comprising the CD blocks of Purba Barddhaman and mostly the western part of the Hooghly district in all four seasons (Fig. 6.2). Additionally, the areal extension of the very high, high, and moderate zone of GWE is large in the monsoon and post-monsoon *rabi* seasons.

Apart from that, the low zone of GWE is located in major part of the study area, especially in the middle east and southern part of the study area. The spatial distribution also depicts the elevation of the groundwater sloping towards the eastern and southeastern parts of the study area (Fig. 6.2). Therefore, the results illustrated that the flow of groundwater is found from the western part of the study area to the eastern and southeastern parts of the study area. This clearly depicts that the sedimentological and other geological conditions support (1) groundwater recharge and (2) diversion of excess monsoon floodwater.

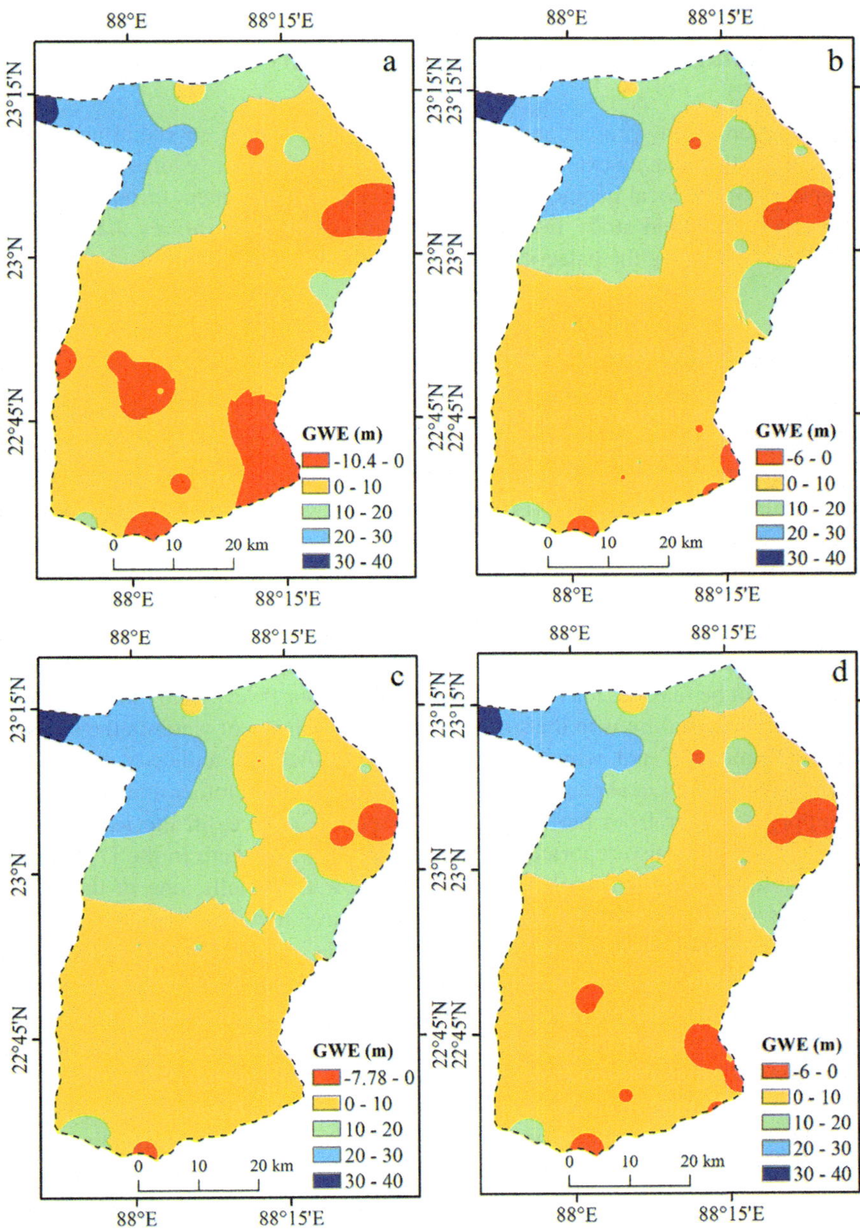

Fig. 6.2 Spatial distribution of average groundwater elevation from 2010–2011 to 2020–2021: (**a**) Pre-monsoon, (**b**) monsoon, and (**c**) post-monsoon *rabi* and (**d**) post-monsoon *kharif* seasons

6.2.2 Socio-Economic Consideration of Water Resource Management Through Palaeochannels and Others

The execution of any project and development plan depends on social needs and demands. To this end, an extensive field survey ($n = 783$) was executed to know the perceptions of the local people about water resource management through the palaeochannel. The field study indicated some important observations that may be helpful for translating the palaeochannel potential into actual reality.

6.2.2.1 Present Nature of Water Supply

Presently, in the study area, most people depend on the water from 'Swajaldhara', a Goverment of India scheme launched on 25 December 2002 under the category of Spot Source Water Supply Schemes (SSWSS) for the supply of safe and clean water to rural people (Roy, 2018). When the project was installed, every household paid INR 1000 is 10% capital cost of the project and every month households need to pay INR 60 as 100% cost of function, maintenance, and management purposes (Roy, 2018). As per the opinions of the local people, the quality of water is good with a regular supply. However, this 'Swajaldhara' scheme is based on the deep tube well extracting huge groundwater. Thus, in the future, the extraction of groundwater may not be feasible for the rapid decline in the water level and the emergence of semi-critical conditions in the study area. Another option of water in the study is from the Damodar Canal, which was constructed by the DVC authority in 1948 for irrigation purposes. Local people use this water for washing clothes and cattle when the supply of water from the 'Swajaldhara' scheme reduces in the lean months (March–May). It is also reported that sometimes they take bath in the DVC canal and get infected with many skin diseases. Shallow tube wells, one of the major sources of water for households and irrigation have now been almost obsolete because of two main reasons: (1) shallow tube wells are not capable of supplying water in the summer seasons because of the fall in the water level and (2) quality of the water from the shallow tube well was not good for drinking due to health reasons. Moreover, water from the palaeochannels was available in the post-monsoon period due to seepage action from the riverbed. However, during 1970–1980, there was huge sand mining from the riverbed of the DFD palaeochannels. They reported that once they have a regular supply of water from the riverbed but after the extraction of sandy layers (aquifers), the supply of water from the riverbed was cut off. 'Mini' pump extracting groundwater is also notable in the study water and mostly this water is used for irrigation purposes. Last but not the least, *doba* and *sech pukur* are two important sources of water for the present area. These ponds are capturing water from rainfall during the monsoon period or from the irrigation canal of the DVC.

6.2.2.2 Demand for Water

Water is desperately needed in the study area, particularly in the semi-critical blocks, especially during the dry months for a variety of domestic and agricultural uses. The field investigations indicate that as there are many existing plans and water sources, at present, they do not face much difficulty in most of the year except for the lean months. However, they aired their concern about the rapid fall of groundwater levels in the locality. One of the senior respondents (Ramendra Mondal, 63 years) of Berala Abad Para village mentioned that before the 1980s, there was availability of water from the riverbed in the post-monsoon months and shallow tube wells could supply adequate water even in the lean months. The water has dried in the shallow tube wells recently.

However, the water demand has been increasing for various reasons. First, the introduction of *boro* cultivation exacerbated the demand for water for irrigation. In the 1980s and 1990s, the large cultivated area of West Bengal comprises rapid growth of boro cultivation due to less production cost and high productivity (Mukherji et al., 2012). Boro rice is cultivated on low land and requires more irrigated water from November–May. Second, an increase in the per capita income led to economic affluence increasing the demand for some purposes like gardening, washing roads, buildings, and bikes and cars. Third, the introduction of bathrooms and toilets instead of community bathing systems and open defecations has increased the demand for water for this. Fourth, mushrooming of cold storage for potato storage and rice mill and other industrial expansions with urban culture also led to huge consumption of water (Dey & Mistri, 2022). The demand for water resources has been shown in Table 6.4.

Table 6.4 Demand of water for domestic purposes (L/cap/day)

Villages	Drinking	Cooking	Bathing	Cloth washing	Domestic animal	Toilet and sanitation	Total
Dhunai	3.02	7.68	22.20	7.32	4.00	12.01	56.23
Kuchut	2.91	9.39	16.74	6.15	0.30	14.16	49.65
Chakbalaram	2.93	6.84	8.89	3.92	4.16	13.25	39.99
Belun	3.16	8.21	12.02	6.39	0.69	11.34	41.82
Berala	3.01	6.84	15.51	5.77	1.14	7.62	39.90
Haral	3.45	7.52	18.48	6.91	0.18	10.46	47.00
Purusattampur	2.58	6.95	16.09	8.78	1.45	11.97	47.81
Bamnagar	3.28	15.56	23.17	7.66	1.91	25.97	77.56
Hazipur	3.46	13.01	22.78	8.79	1.74	22.42	72.20
Morhal	3.55	12.93	29.92	12.58	5.37	18.57	82.91

Sources: Computed from field survey by the authors, 2022

6.2.2.3 Management of Water Demand on a Self-Help Basis

The increasing nature of water demand is being managed by the local community to some extent. For example, people try their best to use minimum water during the lean months. They curtail their water use by 30–40% during the summer season. They used to take baths in the pond instead of the bathroom when necessary. They reuse the water in some cases. For example, cattle feeding with the water already used during washing the vegetables is notable in the area. Almost all households maintain a pond to capture rainwater or DVC canal water for use in the lean months. These clearly indicate that local people are well aware of the water-stressed condition of the DFD and hence they have started adjusting to the existing system. However, they have some demands for more fruitful conservation of water resources. For example, they need more regular supply in the DVC canal instead of its provision in certain phases. As the area is palaeochannel enriched, they have already moulded the palaeochannel into their own ponds to store the monsoon excess water. However, as this pond is small in size and capacity to store water is minimum, they urged for constructing large ponds with the government initiates. They have also expressed their willingness to pay the initial cost of constructing the ponds. For example, 80% of the respondents are ready to pay INR 5000 as one time cost. Moreover, local people are also eager to help the government with the land-acquiring process for community pond generation. They are keener to get the excess supply of floodwater passing through the DVC canal during the monsoon months in their *dobas*, *sech ponds*, and government community ponds (to be constructed). They also expressed their readiness to reduce the *boro* cultivations and switch over to less water-intensive crops and palaeochannel fishing.

In addition, perception analysis regarding the implication of palaeochannels in water resource management has been conducted based on the views of stockholders including senior citizens (age > 60 years), women, planners (Gram Panchayat officials), and school teachers in the study area (Fig. 6.3). The analysis depicts that the majority of the respondent agrees with the proper management of palaeochannels to increase groundwater availability by demining flood events.

6.2.3 Proposed Framework for Integrated Water Resource Management

Reviewing the nature of the geo-physical attributes of the DFD and field survey with the participant's observations depicts that water management in the study area could be done in an integrated manner using some strategies emanating from this investigation. These are stated in the following sections.

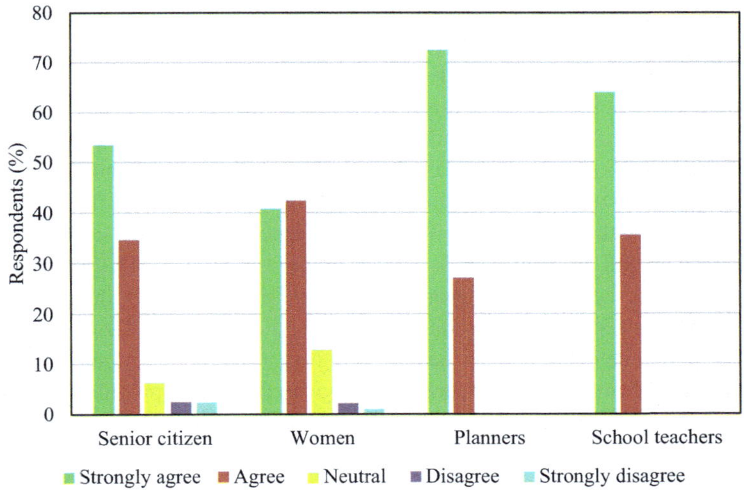

Fig. 6.3 Perception of stockholders regarding water resource management through palaeochannels

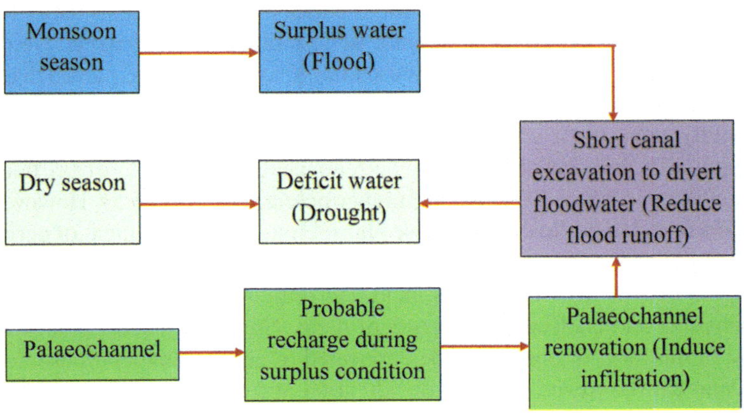

Fig. 6.4 Generalized Framework for integrated water resource management

6.2.3.1 Palaeochannels as Diversion Canal

Palaeochannels are widely distributed in the DFD as discussed in the previous section. Moreover, the scenario of flood and drought is also depicted in the previous section. Naturally, integration between flood, drought, and palaeochannels could be materialized for better water resource management. In this study, flood-prone areas are located predominantly in the southern part of the DFD, while drought-prone areas are mainly found in the northern part of the DFD (Fig. 6.4). Observing the potentiality of the palaeochannels, it is proposed that the high potential area of the DFD is located on palaeochannel 3 and 4 (Fig. 6.1). The northern part (water deficit semi-critical area) could be connected with the present course of the Damodar River, especially during the monsoon season. It is reported that every year during the monsoon period excess floodwater passes through Damodar to inundate its lower portion (Singh et al., 2020). It is pertinent to mention here that there is a debt between arithmetic hydrology and ecohydrology. Arithmetic hydrology believes that the excess floodwater is not directly in the utilization of the ecological and socioeconomic system, while in ecohydrology there is no concept of excess or deficit in a system, rather extremities in the hydrological system are favoured for ecological survival and proliferation. However, in the context of DFD, the acute scarcity of water in one region and devastating flood in another region are strikingly observed. Moreover, if a certain share of water could be diverted from the Damodar River through palaeochannels in the deficit region, it would not hamper the ecological perspective completely. Concerning this view, this study postulates renovations of the palaeochannels to connect the Damodar River with the Hooghly River. It is also proposed that palaeochannel will act as the natural flood diversion conduit. This proposed scheme may benefit the water stress area on one hand and reduce the flood of the lower Damodar River on the other. There is enough scope to renovate the palaeochannel in the study area. The field survey indicates that 100 days' work under the Mahatma Gandhi National Rural Employment Guarantee Act (Mahatma Gandhi NREGA) already excavated some palaeochannels. It is arguably true that excavating long palaeochannels may not be feasible from the perspective of cost–benefit analysis. Thus, this study proposes the renovation of the palaeochannels in some short stretches, for example, renovation of palaeochannel 1 by 5.5 km may be successfully done to divert the excess floodwater to the high recharge potential zone. Similarly, palaeochannel 4 may be renovated by 3.5 km to divert floodwater to the middle part of the DFD having higher potentiality (Fig. 6.5). However, there might be a few changes to execute these plans. First, either settlement or agricultural lands occupy most of the palaeochannels area. Table 6.5 depicts that the major land use and land cover (LULC) classes including settlement have developed along the palaeochannel and enclosed some parts of them. However, narrow palaeochannels existing in the form of ponds can be connected to get the flow continuity of water from Damodar. However, careful consideration and an in-depth diagnostic survey are required before implementing this plan. Second, in some cases, some agricultural plots may be shifted from the core of the palaeochannels (Fig. 6.5). Moreover, other infrastructural developments like road, cold store, and rice mill, which are located in the vicinity of palaeochannel require a careful management plan. For example, the crossing of palaeochannels and roads may necessitate the construction of bridges and culverts. Considering these challenges, this study proposes a small excavation instead of a long excavation. However, future planners and stockholders may think of the Damodar to Hooghly connection via palaeochannels considering the infrastructural inertia.

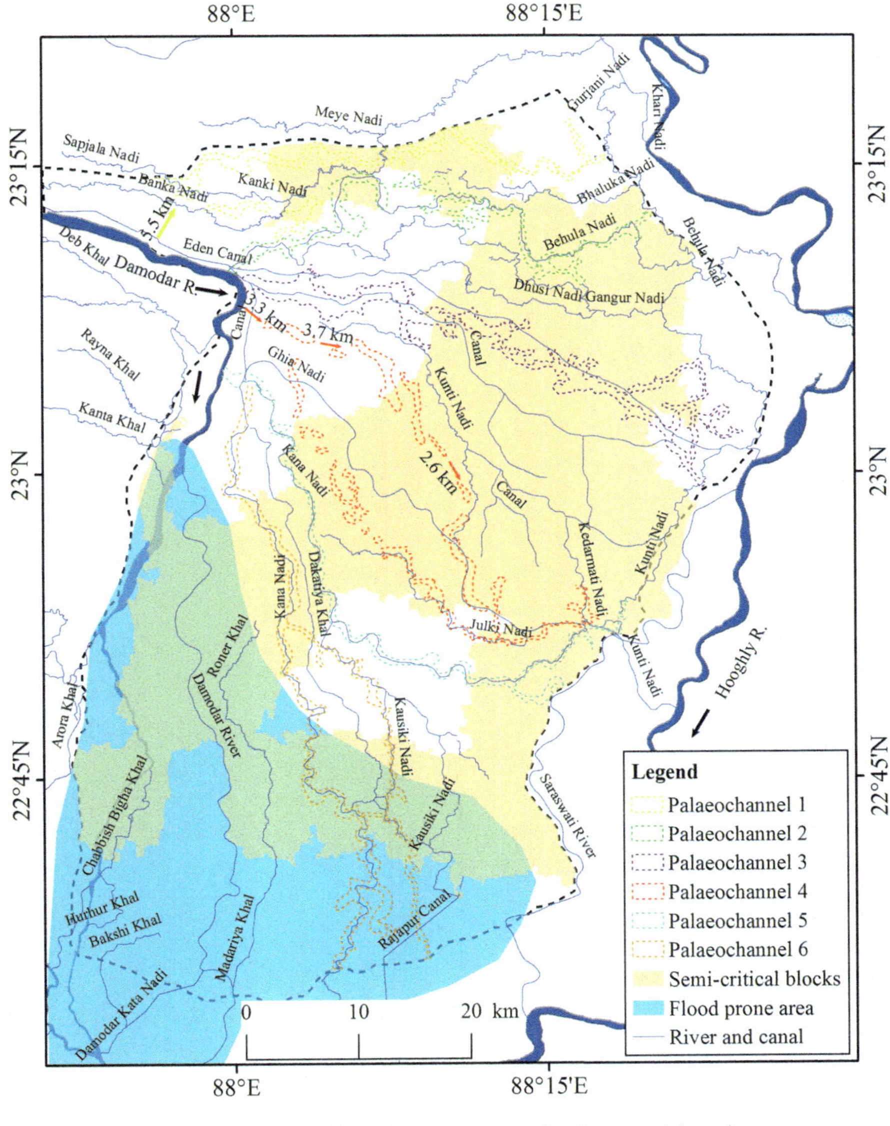

Fig. 6.5 Renovation of the palaeochannels to connect the flood area and drought area

6.2.3.2 Palaeochannels as Individual and Community Ponds

Palaeochannels could also be potentially developed as ponds managed by individual and community initiatives. In the context of DFD, it is observed that in the water-stressed areas, local people are already using small ponds (~100 m^2) locally called *doba* for capturing rainwater during the monsoon months or excess water from DVC canals. Figure 6.6 portrays that the practice of rainwater harvesting on a

Table 6.5 Properties of the major palaeochannels

Palaeochannel name	Length (km)	No. of major road crossing	No. of railway crossing	Occupied by major types of LULC (%)					
				Waterbody	Vegetation cover	Agricultural land	Agricultural fallow land	Sandy area	Built-up area
Palaeochannel 1	54	2	0	6.39	21.47	34.97	31.67	0.30	5.20
Palaeochannel 2	54.2	4	1	5.35	23.08	27.28	38.83	0.99	4.47
Palaeochannel 3	50	10	4	7.79	27.26	33.54	24.50	0.68	6.22
Palaeochannel 4	56.5	5	1	5.20	29.83	39.15	22.10	0.14	3.58
Palaeochannel 5	73.9	5	2	5.99	32.01	29.84	26.84	0.84	4.48
Palaeochannel 6	84.6	7	2	6.37	35.79	33.09	19.52	0.75	4.48

Sources: Computed by the authors (2022)

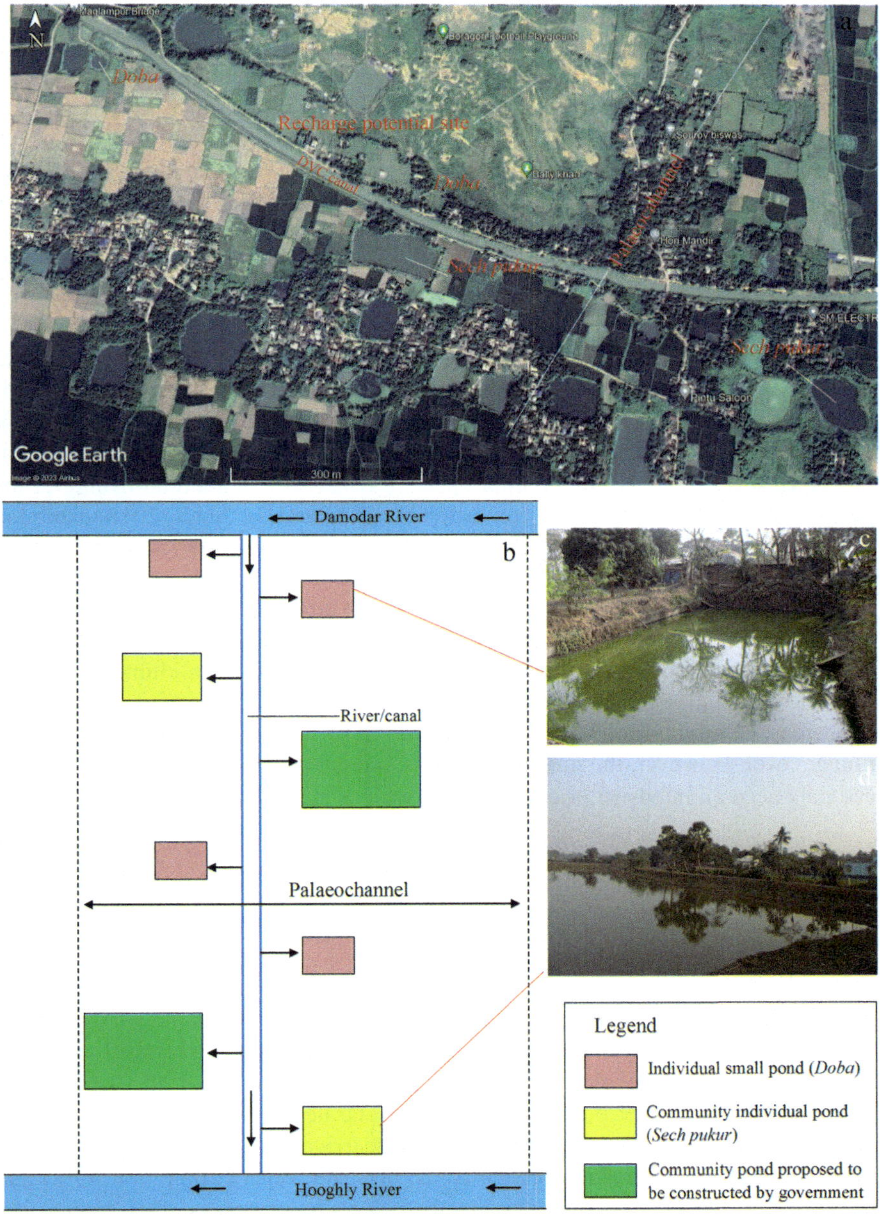

Fig. 6.6 Schematic plan for water supply in the lean month, groundwater recharge, and fishing: (**a**) Distribution of surface water bodies in Google Earth image, (**b**) plan for water storage in individual, community, and government constructed ponds, (**c**) small community pond or *doba*, and (**d**) community individual pond or *sech pukur*

rudimentary basis is observed. Two types of ponds are maintained by the local people. The first one is the *doba* or individual pond managed at the household level, while *sech pukur* (~100,000 m^2) are maintained at the community level. The field survey indicates that local people are well aware of the surplus water passing through the DVC canal during monsoon. Interaction with the household members clearly indicates their inclination to use this access floodwater in their ponds. They also mention that the government may construct community ponds of large size because there is availability of potential land for ponds. If this pond could be potentially developed, it would fulfil a few major objectives: (1) This pond will store excess water that may be used for the agricultural field and other purposes during the lean seasons. (2) These ponds will recharge groundwater because most of the ponds are located within the potential recharge zone. Another important point that the local people mentioned was the potential fishing in the palaeochannels converted community ponds. The survey at the household level revealed that about 70–80% of respondents are eager to be engaged in palaeochannel fishing. The villagers irrespective of fishing communities are eager for commercial fishing in the palaeochannels. A similar study conducted by Islam and Guchhait (2018) exhibited palaeochannel fishing in the Bhagirathi river basin to better cope with bank erosion hazards. Further, another study conducted by Islam et al. (2022) made an estimate of income and expenditure from palaeochannel fishing and the number of families to be supported by minimum support income. Moreover, maintenance of water in the pond throughout the year may have a stimulating effect on climate soothing. Considering this perspective, this study proposed construction of a government-managed community pond and the maintenance of existing individual pons at the village level. However, the number of ponds and the exact location with budgetary estimate need an intensive investigation by various stakeholders.

6.3 Conclusions

Palaeochannel plays a vital role in water resources management. This chapter has outlined the role of the palaeochannel in the quantitative and qualitative augmentation of both the surface and groundwater. Groundwater depth is higher in the non-palaeochannel wells compared to the palaeochannel wells of the study area. Moreover, surface water bodies are more concentrated in the areas adjacent to the palaeochannel than those areas located far away. Irrigation and drinking water qualities are better for the palaeochannel areas; however, organic pollution (OIP) and heavy metals are more concentrated in the palaeochannel areas. This chapter has demonstrated the scope of floodwater supply in drought-prone areas in a reasonable manner and the roles of the palaeochannel in floodwater distribution. In general, the northern part is drought-prone, while the southern part of the DFD is flood-prone. Observing these hydrological extremities in the study area, geophysical conditions of the palaeochannel for recharge potential, and socio-economic consideration, the study proposed a two-component framework for water resources management. The first is related to the potential of the palaeochannel in the distribution of floodwater from surplus to deficit areas. The second component considers the palaeochannel for individual and community ponds for groundwater recharge and palaeochannel fishing. These models may play a crucial role for both flood- and drought-prone areas for future development.

References

Aslam, M. A., Balasubramanian, A., Kondoh, A., Rokhmatuloh, & Mustafa, A. J. (2003). Hydrogeomorphological mapping using remote sensing techniques for water resource management around palaeochannels. *IGARSS 2003. 2003 IEEE International Geoscience and Remote Sensing Symposium. Proceedings (IEEE Cat. No.03CH37477).* IEEE. doi:https://doi.org/10.1109/igarss.2003.1294768.

CGWB. (2014). *Dynamic ground water resources of India (as on 31st march 2011).* Ministry of Water Resources, River Development & Ganga Rejuvenation Government of India.

Chaudhary, B. S., & Aggarwal, S. (2009). Demarcation of palaeochannels and integrated ground water resources mapping in parts of Hisar district, Haryana. *Journal of the Indian Society of Remote Sensing, 37*, 251–260. https://doi.org/10.1007/s12524-009-0019-5

Dey, C. K., & Mistri, T. (2022). Spatio-temporal changes of crop combination in selected C.D. Blocks of Purba Bardhaman District, West Bengal, India. In *Livelihood enhancement through agriculture, tourism and health* (pp. 185–203). Springer Nature Singapore. https://doi.org/10.1007/978-981-16-7310-8_9

Islam, A., & Guchhait, S. K. (2018). Analysis of social and psychological terrain of bank erosion victims: A study along the Bhagirathi River, West Bengal, India. *Chinese Geographical Science, 28*, 1009–1026. https://doi.org/10.1007/s11769-018-0937-7

Islam, A., Ghosh, S., Barman, S. D., Nandy, S., & Sarkar, B. (2022). Role of in-situ and ex-situ livelihood strategies for flood risk reduction: Evidence from the Mayurakshi River Basin, India. *International Journal of Disaster Risk Reduction, 70*, 102775. https://doi.org/10.1016/j.ijdrr.2021.102775

Johnson, A. I. (1967). Specific yield: compilation of specific yields for various materials. *U. S. Geological Survey Water-Supply Paper* 1662-D, 74.

Kakade, A., Salama, E.-S., Han, H., Zheng, Y., Kulshrestha, S., Jalalah, M., et al. (2021). World eutrophic pollution of lake and river: Biotreatment potential and future perspectives. *Environmental Technology & Innovation, 23*, 101604. https://doi.org/10.1016/j.eti.2021.101604

Kaur, A., Bhatnagar, S., Krishan, G., & Rana, R. (2024). Hydrogeophysical mapping of paleochannels for water security in Bhawanigarh Block, District Sangrur, Punjab, India. *Groundwater for Sustainable Development, 27*, 101354.

Mahammad, S., & Islam, A. (2021). Evaluating the groundwater quality of Damodar Fan Delta (India) using fuzzy-AHP MCDM technique. *Applied Water Science, 11*. https://doi.org/10.1007/s13201-021-01408-2

Ministry of Water Resources Government of India. (2009). *Report of the ground water resource estimation committee groundwater resource estimation methodology.* Ministry of Water Resources.

Mukherji, A., Shah, T., & Banerjee, P. S. (2012). Kick-starting a second green revolution in Bengal. *Economic and Political Weekly*, 27–30.

Pal, T., & Mukherjee, P. K. (2008). Study of subsurface geology in locating arsenic-free groundwater in Bengal delta, West Bengal, India. *Environmental Geology, 56*, 1211–1225. https://doi.org/10.1007/s00254-008-1221-4

Pal, T., & Mukherjee, P. K. (2010). Search for groundwater arsenic in Pleistocene sequence of the Damodar River flood plain, West Bengal. *Indian Journal Of Geosciences, 64*, 109–112.

Rathore, V. S., Nathawat, M. S., & Champatiray, P. K. (2010). Palaeochannel detection and aquifer performance assessment in Mendha River catchment, Western India. *Journal of Hydrology, 395*, 216–225. https://doi.org/10.1016/j.jhydrol.2010.10.026

Roy, A. (2018). Impact of Swajaldhara scheme on the rural livelihoods: A case study of Purba Medinipur district in West Bengal. *Vidyasagar University Journal of Commerce, 23*. https://ir.vidyasagar.ac.in/handle/123456789/5126

Samadder, R. K., Kumar, S., & Gupta, R. P. (2011). Paleochannels and their potential for artificial groundwater recharge in the western Ganga plains. *Journal of Hydrology, 400*, 154–164. https://doi.org/10.1016/j.jhydrol.2011.01.039

Sarkar, B., & Islam, A. (2019). Drivers of water pollution and evaluating its ecological stress with special reference to macrovertebrates (fish community structure): A case of Churni River, India. *Environmental Monitoring and Assessment, 192*. https://doi.org/10.1007/s10661-019-7988-9

Singh, R. K., Villuri, V. G., Pasupuleti, S., & Nune, R. (2020). Hydrodynamic modeling for identifying flood vulnerability zones in lower Damodar river of eastern India. *Ain Shams Engineering Journal, 11*, 1035–1046. https://doi.org/10.1016/j.asej.2020.01.011

Sinha, R., Yadav, G. S., Gupta, S., Singh, A., & Lahiri, S. K. (2013). Geo-electric resistivity evidence for subsurface palaeochannel systems adjacent to Harappan sites in Northwest India. *Quaternary International, 308-309*, 66–75. https://doi.org/10.1016/j.quaint.2012.08.002

Chapter 7
Conclusions

7.1 Major Findings

7.1.1 Palaeochannel Detection

- The identification of palaeochannels has been done using optical and microwave remote sensing data processing and geophysical data analysis, which is closely associated with the literary evidence and historical records.
- The important place locations on the Gangur, Behula, Bimala, and Brahmani rivers during the medieval period as mentioned in the *Manasa Mangal* and *Dharma Mangal Kavya* closely correspond to the present-day locations of the palaeochannels of the Damodar River.
- The historical maps recording the spatial distribution of rivers provide useful information on the ancient river courses. The historical maps also help detect the sequential changes in the course of the Damodar River, leading to the confirmation of the presence of palaeochannels.
- This investigation, primarily based on low-cost palaeochannel detection techniques using satellite images, has detected DFD palaeochannels that are similar to the abandoned channels of the present river. The effectiveness of the study reveals that the optical data provides better information related to surface configurations, while the SAR data portrays the near-surface features of palaeochannel in a better way. Besides, the width of palaeochannels is measured by the surface profile, which reveals the accuracy of the optical and radar data. The validation of the study demonstrated that the data and methods used were appropriate to demarcate the palaeochannels and reveal the past fluvial system.
- Moreover, lithological layers indicate that the palaeochannels of the study area are associated with the combination of coarse materials such as *kankar*, pebbles, and gravel with clay and sand, which indicates the high kinetic energy of the

S. Mahammad, A. Islam, *Palaeochannels and Water Resource Management*,
https://doi.org/10.1007/978-3-031-92021-9_7

riverine system. The presence of coarse material is found along the demarcated palaeochannels by remote sensing data, which also confirms the existence of palaeochannels.

7.1.2 Groundwater Dynamics and Quality Assessment

- The temporal variation of GWL (measured in mbgl) in the study implies that except for a few locations, most of the wells have an increasing trend in the groundwater depth implying depletion of groundwater resources. The seasonal distribution of GWL revealed that the maximum increase in GWL occurred in the post-monsoon *kharif* season followed by pre-monsoon, monsoon, and post-monsoon *rabi* seasons.
- Out of the six machine learning algorithms, the XGB regressor became the most reliable model for future prediction, while the DT proved the least efficient for portraying the GWL dynamics. The predicted GWL of 2025–2026 also indicates the increasing trend in groundwater depth for most of the wells, especially in the western part of the DFD around Barddhaman town.
- The spatiotemporal variations of GWL are significantly related to rainfall and population dynamics in the study area. This implies that an increase in groundwater depth is associated with a decrease in rainfall. Moreover, an increase in population, urbanization, industrialization, and agricultural expansions led to the overextraction of groundwater resources for irrigation, domestic, and industrial sectors causing an increase in the number of semi-critical blocks in the DFD.
- Along with the depletion of the groundwater level, this inquiry has also attempted to evaluate the groundwater quality for drinking and irrigation use. The WQI shows that 50% of the sample are excellent, while 11.76% of the samples are very poor. The results also show 17.65% of the samples are of very good quality and 20.59% of the samples are of poor water quality for drinking purposes. The EIWQI of groundwater shows that 52.94%, 41.18%, 2.94%, and 2.94% of the sampled wells are recorded as excellent, good, moderate, and poor, respectively, in terms of the quality of irrigation water.
- The spatial distribution of the EIWQI of groundwater reveals that most of the samples of this study area are of excellent and very good quality and distributed in the eastern, western, and southern fringe areas. The spatial pattern of the EIWQI appears similar to the alignment with the spatial variation of WQI. The result revealed that EC, TDS, Cl^-, Na^+, and HCO_3^- were the most significant parameters determining the quality of groundwater. A high concentration of EC, Cl^-, and Na^+ results from the increasing salinity due to high evaporation and anthropogenic activities.

7.1.3 Surface Water Dynamics and Quality Assessment

- The supervised LULC classification has been implemented to detect the changes in major land categories, such as waterbody, vegetation cover, agricultural land, agricultural fallow land, sand area, and built-up area of 1989, 1999, 2010, and 2020 both pre-monsoon and post-monsoon seasons. In this regard, widely used machine learning (ML) algorithms such as random forest (RF), support vector machine (SVR), and classification and regression tree (CART) have been applied to the Google Earth Engine. The results depict that waterbody in the pre-monsoon period increased by 5.53 km^2 (0.17%) from 1989 to 2020. Moreover, in the post-monsoon season, the waterbody class covers 110.44 km^2 in 1989, whereas it comprises 96.76 km^2 in 2020, which reveals a declining trend of waterbody −13.68 km^2 (−0.43%) from 1989 to 2020. Therefore, the dynamics of waterbodies are closely associated with the rainfall of the POM and POM kharif seasons in the study area as discussed in Chap. 4.
- As per LULC classification, the loss of vegetation has been found in both the pre-monsoon (−0.32%) and the post-monsoon periods (−5.69%). Besides, agricultural land has recorded significant growth in pre-monsoon and post-monsoon seasons, with increases of +28.7% and + 15.84%, respectively, during the same period. Also, agricultural fallow land has been observed as a decreasing land category in pre-monsoon (−30.55%) and post-monsoon season (−13.15%) from 1989 to 2020. Moreover, in the pre-monsoon and post-monsoon periods, the increasing trend of the built-up area has been observed at +2.19% and +2.59%, respectively.
- In addition, intensity analysis has been done in terms of interval intensity, categorical intensity, and transition intensity to detect the dynamics of LULC in the study area. The interval level intensity depicts that the change rate of the time interval of 1989–1999 has been categorized as fast change by crossing the uniform intensity line for both pre-monsoon and post-monsoon seasons. Besides, in the category level intensity, vegetation cover, agricultural land, and agricultural fallow land have been considered as the active loser and gainer land class in pre-monsoon and post-monsoon seasons through the study intervals.
- Transition level analysis has suggested that vegetation cover, agricultural land, agricultural fallow land, and the built-up area largely control the dynamics of the LULC of the study area. The accuracy assessment of the ML algorithms reveals that the random forest classifier scored highest in terms of overall accuracy and Kappa coefficient of all the seasons indicating the best-fit classifier for image classification. Therefore, the supervised image classifications of all seasons from 1989 to 2020 have been done using an RF classifier.
- It is observed that the seasonal distribution of rainfall is closely associated with the distribution of the waterbodies in the study area. The reduction of vegetation cover has occurred through the expansion of the built-up area and agricultural

land in the study area. Besides, the growth of vegetation cover in a few areas depicts the transformation of agricultural fallow and cropland into orchard cultivation and plantation. Additionally, agricultural land and built-up area have been increased due to increasing population and socio-economic growth.

- The WQI of surface water shows that 19.44% of the sample are excellent, 33.33% are good, 30.56% are moderate, 11.11% are poor, and 5.55% of the samples are very poor in terms of drinking purposes. The EIWQI reveals that 19.44%, 47.22%, 30.55%, and 2.78% of the sample wells are considered excellent, good, moderate, and poor, respectively, in terms of the quality of irrigation water.
- Moreover, the spatial distribution of EIWQI of surface water shows that unsuitable water quality for irrigation is concentrated in the eastern part of the study area, whereas a good category of EIWQI is found in the middle portion and western and eastern fringe of the study area.
- In addition, only two sample locations (~8%) are recorded in the C1 class, indicating excellent condition and the rest of the samples (92%) are located in the C2 class indicating the acceptance of the surface water quality. The major pollution sources identified in the study area are agricultural wastage, organic pollution, and influences of anthropogenic activities.

7.1.4 Palaeochannels in Water Resource Management

- Spatial distribution of flood inundation areas and drought-affected areas has been detected for knowing the surplus deficit condition in the DFD required for water resource management. In the study area, flood conditions of 2015, 2016, 2017, and 2019 have been assessed using Sentinel 1A SAR data. The results depict that the major flood inundation is mainly concentrated in the southern part of the DFD.
- The drought condition of the study area has been detected for 2010 and 2020 in pre-monsoon, monsoon, post-monsoon *rabi*, and post-monsoon *kharif* seasons. GWDI infers that severe to extreme drought has occurred in the monsoon and post-monsoon *rabi* seasons in 2010 and 2020. The spatial distribution of GWDI shows that the severe to extreme drought is located in the central-eastern part of the study area.
- Palaeochannel plays a crucial role in the management of surface and groundwater resources. The study findings indicate that groundwater depth is lower in the palaeochannel area compared to the non-palaeochannels area. Moreover, groundwater quality in terms of irrigation and drinking purpose is good in palaeochannel areas in comparison to the non-palaeochannels areas. Similarly, the extension of the surface waterbody is more in case of the palaeochannels area. Moreover, some parameters of drinking water quality are good in the palaeochannels area.

- Observing the recharge potentiality of the palaeochannel and socio-economic demands for water in the water-stressed area, this study has proposed an integrated water resource management using palaeochannels. This model has two components: (1) palaeochannels as flood diversion canals for distributing excess flood water into water-stressed areas and (2) construction of individual and community ponds for groundwater recharge and palaeochannel fishing.

7.2 Limitations and Future Scope of Research

After carrying out the entire work, this study has faced some problems and limitations in different aspects of palaeochannel identification and water resource management. The first limitation is concerned with the delineation of palaeochannels in some areas where the palaeochannels have lost their existence due to the expansion of agricultural land, settlements, and the excavation of canals and ponds along the palaeochannel. However, this problem is to some extent overcome using the optical, microwave, literary, and historical records and lithological evidence. The second limitation indicates that the groundwater system is much more complicated in alluvium rather than in the hard rock hydrogeological setup. In alluvium, groundwater belongs to the mostly unconfined aquifer and sometimes it is occupied in a leaky and confined aquifer. Clay behaves like an aquitard between the vertical sequences of the aquifer. The third relates to the lesser availability of static groundwater level data which is problematic in this study. The depth of groundwater resources of the different blocks is very limited in a few sites. Finally, water resource management using palaeochannels may have a few challenges, e.g. renovation of palaeochannels to transform them into flood diversion conduits and canals and acquisitions of land for constructing community ponds.

This study has several important observations as outlined at the very outset of this chapter. However, observing the limitations of this study, it is felt that some less-explored areas of this research could potentially be examined by future organizations and researchers. For example, the delineated palaeochannels are distributed in a radial pattern and bifurcated in their lower courses, which indicate the process of the delta formation, tectonic movements, and sea-level changes in the study area. Further study should be taken in this direction. Besides, as the deeply buried portion of the palaeochannel is quite difficult to delineate using remote sensing image processing due to its incapability to detect the features concealed at greater depth under the surface layer, the future discourse may focus on the application of intensive geophysical survey coupled with advanced geospatial data and techniques. Finally, the proposed framework of integrated water resource management needs to be further introspected with more detailed socio-economic surveys and the perception of the various stakeholders and planners in similar regions of the world.

Glossary

Aman A variety of rice is sown in the rainy season and harvested in winter
Aus A variety of rice is sown in summer with the pre-monsoonal showers and harvested in autumn
Boro A variety of rice is sown in winter and harvested in summer
Dharmamangal Kavya The most significant piece of medieval Bengali literature and an important subgenre of *Mangalkavya* that narrates of local deities of rural Bengal
Doba One kind of pond/small waterbody
Fan Delta A fan delta is a depositional feature that is formed where an alluvial fan develops directly in a body of standing water from some adjacent highland
Kana River Dried or lost river
Kharif season Summer season where crops are sown at the beginning of the summer rains
Khesari Indian pea or blue sweet pea
Manasa Mangal kavya The oldest of the Mangal Kavya and narrates how Manasa, the snake-goddess, introduced her worship in Bengal
Musuri Red or orange lentils—a variety of Indian pulses
Nadi River-channelized flow of water
Rabi season Winter season where crops are sown in winter and harvested in the spring in India
Sech pukur A pond mainly used for irrigation
Swajaldhara A rural drinking water project for establishing water supply systems

S. Mahammad, A. Islam, *Palaeochannels and Water Resource Management*,
https://doi.org/10.1007/978-3-031-92021-9

Index

S. Mahammad, A. Islam, *Palaeochannels and Water Resource Management*,
https://doi.org/10.1007/978-3-031-92021-9

The manufacturer's authorised representative in the EU is Springer Nature Customer Service Centre GmbH, Europaplatz 3, 69115 Heidelberg, Germany. If you have any concerns regarding our products, please contact ProductSafety@springernature.com

Printed and bound by CPI Group (UK) Ltd, Croydon, CR0 4YY
01/10/2026
02225883-0017